Corrosion and its Consequences for

Reinforced Concrete Structures

Structures Durability in Civil Engineering Set

coordinated by
Christian La Borderie and Alain Sellier

Corrosion and its Consequences for Reinforced Concrete Structures

Raoul François
Stéphane Laurens
Fabrice Deby

First published 2018 in Great Britain and the United States by ISTE Press Ltd and Elsevier Ltd

ISTE Press Ltd
27-37 St George's Road
London SW19 4EU
UK

www.iste.co.uk

Elsevier Ltd
The Boulevard, Langford Lane
Kidlington, Oxford, OX5 1GB
UK

www.elsevier.com

For information on all our publications visit our website at http://store.elsevier.com/

British Library Cataloguing-in-Publication Data
A CIP record for this book is available from the British Library
Library of Congress Cataloging in Publication Data
A catalog record for this book is available from the Library of Congress
ISBN 978-1-78548-234-2

Printed and bound in the UK and US

Contents

Introduction

This book focuses on the corrosion of reinforcements in reinforced concrete. It is more specifically aimed at Civil Engineers faced with the issue of aging in structures affected by corrosion.

As such, the mechanisms of steel corrosion in reinforced concrete are presented here, while recalling the basics of electrochemistry and placing particular emphasis on the specific nature of corrosion in reinforced concrete: its localized character.

We then develop on the consequences of the large dimensions of Civil Engineering structures and their operation: the presence of defects in the casting of fresh concrete, the existence of service cracks and mechanical damage at the steel-concrete interface. The respective roles of these elements are clarified.

Corrosion quantification is then presented. To this end, two parameters have been selected: local cross-section (LC) loss and generalized cross-section (GC) loss, calculated based on the space between service cracks or shear stirrups.

The corrosion diagnosis is then discussed: the only quantitative measure available today is the recording of the lengths and openings of cracks resulting from the corrosion process. This corrosion signature is used to establish a diagnosis of the local cross-section (LC) and generalized cross-section (LC) losses throughout this book.

The mechanical consequences of reinforcement corrosion are then studied, in particular the change in the constitutive equation of steel behavior after corrosion, with it becoming more fragile. Changes in bearing capacity, bending stiffness and ultimate deflection at failure caused by corrosion are presented.

We then go on to look at the prediction of residual mechanical behavior based on the data of a visual diagnosis, discussing modes of failure, bearing capacity, ultimate deflection, the formation capacity of plastic hinges and the calculation of deflections in service of corroded elements.

Chapter 7 provides indications as to how to predict the service life of structures in the presence of corrosion.

The final section presents the repair possibilities enabling corrosion to be reduced: cathodic protection and electrochemical treatments of chloride-contaminated or carbonated structures.

1

Steel Corrosion in Reinforced Concrete

1.1. Introduction

Steel corrosion constitutes the most important cause of premature aging and deterioration on reinforced-concrete structures on an international scale. Its technical, financial and societal consequences are considerable. Its economic impact in particular is hard to assess as it would require incorporation of both direct costs (surveillance, maintenance, repairs and rehabilitation) and indirect costs (operating losses relating to immobilization of the structure, insurance, etc.) to define a relevant measurement of the cost relative to steel corrosion in concrete. Across all sectors as a whole, certain sources assess the total cost of corrosion at around 4% of GDP on average in industrialized countries. The specific case of steel corrosion in concrete certainly contributes significantly to this percentage and the sums allocated annually to the rehabilitation of corroded reinforced-concrete structures stands at billions of Euro.

Regarding other industrial sectors also affected by this issue, concrete reinforcement corrosion presents specific features that have posed major difficulties for the scientific community, both in terms of phenomenological understanding (initiation conditions, kinetic control mechanisms, service-life prediction, etc.) and in terms of metrology and electrochemical maintenance.

Within the wide field of civil engineering research, concrete reinforcement steel corrosion has represented a major theme for over four decades, originally motivated by the reporting of the first significant damage. In chronological terms, corrosion development processes were first studied in order to identify the main causes and to put forward solutions dedicated to the construction of structures better able to withstand this pathology. Having identified the two main causes (chlorides and/or atmospheric CO_2), the scientific community concerned became aware of the more or less inescapable nature of steel corrosion in concrete, and standardization documents governing construction gradually evolved to include an increase in the concrete cover of steel reinforcements in order to delay its initiation as much as necessary. Research within the field has also explored the impact of the type of cement used in terms of corrosion resistance. These different works, which began in the late 1960s, remain relevant today.

More recently, faced with a proven, widespread problem on a worldwide scale, it became necessary to diagnose a structure's state of corrosion in a non-destructive manner. Thus, intensive research began into electrochemical techniques dedicated to measuring corrosion in reinforced concrete.

Of course, these techniques are directly derived from the measurement tools developed in the electrochemistry field. Generally speaking, research within this particular field has for too long been based on an exclusively experimental, empirical approach of studying the phenomena at stake. Knowledge and practices tried and tested in general laboratory electrochemistry or in other industrial sectors were applied without taking account of the specific characteristics of reinforced concrete. Indeed, few studies went to the effort of identifying the fundamental differences existing between a measurement made in an electrochemistry laboratory on a fully controlled electrode plunged into a calibrated electrolytic solution, and a measurement collected *in situ* on a reinforcement layout subjected to numerous sources of uncertainty and plunged into a complex, changing and uncontrolled electrolytic environment (concrete).

As such, despite thousands of articles and other scientific works on the subject, the question of steel corrosion in reinforced concrete as an engineering problem has still not been resolved. Notably, the lack of reliability in corrosion measurement techniques, in particular its kinetics, is almost unanimously acknowledged. By ranking the causes of this failure, two physical assumptions that may globally be defined as a "paradigm of uniformity" come out on top:

– Assumption regarding the uniformity of the electrochemical state of the reinforcement layout.

– Assumption regarding the uniformity of the electrochemical action with respect to the corrosion rate measurements and electrochemical maintenance techniques.

The development of the paradigm of uniformity was based essentially on the deceptive impression that it could transform a 3D physical problem into an equivalent straightforward 1D problem. Implicitly or explicitly, a great number of current applications dedicated to steel corrosion in concrete are based on this concept of an equivalent 1D problem:

– Linear polarization resistance measurements [AND 04].

– Using the Stern-Geary equation to convert an apparent polarization resistance into a corrosion rate.

– Cathodic protection and the concept of "instant-off" potential used as a performance criterion.

– Correcting ohmic drop via Randles' equivalent electrical circuit model.

For the last 15 or so years, faced with this clear deadlock, a change in paradigm has begun to be applied, at last paving the way for real technical control of the problem. This shift depends on the strict observance of the 3D nature of the physical problem, which cannot be reduced to an equivalent 1D problem: the localized nature of corrosion in reinforced concrete (due to the non-uniform electrochemical state of the reinforcement layout) and non-uniform distribution of polarization currents, whether within the context of corrosion rate measurements, or electrochemical maintenance [SAG 92, ELS 02, WAR 06, LAU 16, ANG 15, MAR 16, SAS 16]. At the present stage, this technical control has not yet been finalized, but it should certainly begin to dominate within the next decade through the development of:

– Robust engineering models and 3D numerical simulation tools.

– More-consistent metrological approaches in line with the phenomenological reality.

– Optimized maintenance and prevention techniques.

1.2. General corrosion theory concepts

This chapter aims to provide a summary definition of the respective notions of a uniform corrosion system and a galvanic (or localized) corrosion) system based on the fundamental concept of reversible electrode.

1.2.1. *Reversible electrode*

A pure, homogeneous metal (M), plunged into an electrolyte containing M^{z+} ions, is subjected to the following equilibrium reaction (equation [1.1]):

$$M \leftrightharpoons M^{z+} + z\,e^- \quad [1.1]$$

At equilibrium, the oxidation rate is equal to the reduction rate. The dissolution reaction is constantly compensated for by the reduction reaction. There is therefore no apparent loss in metallic mass and, consequently, no corrosion. The term "reversible electrode" is then referred to. A reversible equilibrium potential is associated with this state of thermodynamic equilibrium, E_{rev}. The reversible equilibrium potential of an electrode plunged into an electrolyte is defined by the Nernst law.

For a given reversible equilibrium between the reduced (Red) and oxidized (Ox) forms of the same chemical species (equation [1.2]):

$$Red \leftrightharpoons Ox + z\,e^- \quad [1.2]$$

The Nernst law is expressed by equation [1.3]:

$$E = E_0 + \frac{R\,T}{z\,F} \ln\left(\frac{a_{Ox}}{a_{Red}}\right) \quad [1.3]$$

whereby:

– E_0 is the standard potential of the electrode measured with respect to the normal hydrogen electrode [V/NHE];

– R is the gas constant;

– T is the temperature [K];

– F is the Faraday constant [96,500 C];

– a_{Ox} is the activity of the oxidized form;

– a_{Red} is the activity of the reduced form.

Put very simply, standard potential involves measurement of the energy barrier to be crossed in order to oxidize a species. As such, the hierarchy of the standard potentials of reversible electrodes defines their oxidation tendencies. The lower the standard potential, the more the energy barrier is reduced and the more the species is likely to oxidize. As an example, the redox couple for gold (Au/Au^{3+}) presents a standard potential of +1.5 V/NHE, thus explaining its significant stability in its reduced metallic form (Au). Inversely, sodium (Na/Na^{+}) is characterized by a standard potential of −2.7 V/NHE. Hence, sodium is extremely unstable in its metallic state (Na).

At reversible equilibrium (Figure 1.1), the net electric flux apparent at the metal-electrolyte interface is zero as the anodic flow (or current) associated with oxidation is equal to the cathodic flow relative to reduction. Assuming that the phenomena occur in a uniform manner on the metallic surface, it is possible to reason in terms of net current density apparent at the interface (i) (equation [1.4]):

$$i = i_a + i_c \quad [1.4]$$

Whereby i_a is the anodic current density (reflecting the kinetics of the oxidation reaction) and i_c the cathodic current density (reflecting the kinetics of the reduction reaction). By convention, an anodic current is counted positively and a cathodic current is counted negatively. At equilibrium, the anodic and cathodic kinetics are equal in absolute value ($i_a = -i_c$) and the apparent net current density is zero ($i = 0$).

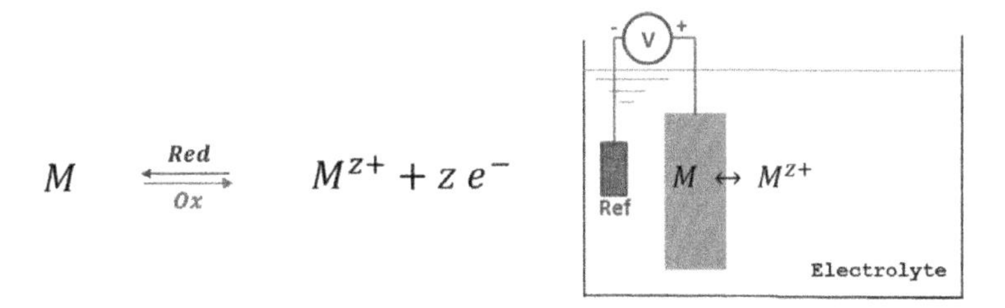

Figure 1.1. *Illustration of the concept of reversible equilibrium. For a color version of the figure, see www.iste.co.uk/francois/corrosion.zip*

With the aid of a simple device, it is possible to unbalance this system in two distinct ways:

– forcing the anodic reaction (oxidation) by “pumping” the electrons from the metal: this is referred to as anodic polarization of the system (Figure 1.2);

– forcing the cathodic reaction (reduction) by supplying electrons to the metal: this is known as cathodic polarization (Figure 1.3).

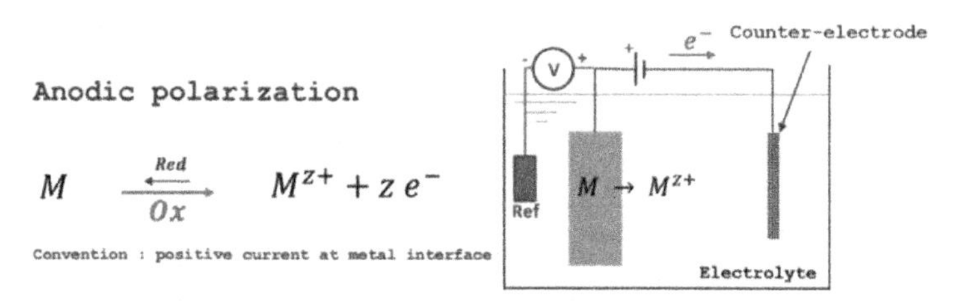

Figure 1.2. *Anodic polarization of a reversible electrode. For a color version of the figure, see www.iste.co.uk/francois/corrosion.zip*

Anodic polarization forces the oxidation reaction and thus causes an unbalance in the currents, revealed by a positive apparent net current at the interface (equation [1.5]):

$$|i_a| > |i_c| \Rightarrow i = i_a + i_c > 0 \qquad [1.5]$$

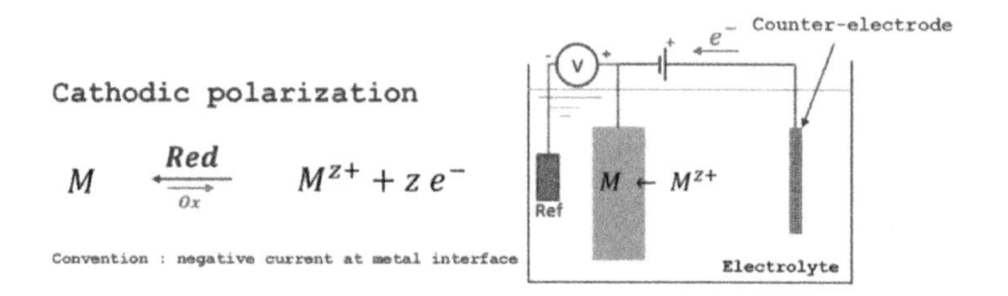

Figure 1.3. *Cathodic polarization of a reversible electrode. For a color version of the figure, see www.iste.co.uk/francois/corrosion.zip*

In cathodic polarization, forcing the reduction reaction is reflected by a negative apparent net current (equation [1.6]):

$$|i_a| < |i_c| \Rightarrow i = i_a + i_c < 0 \qquad [1.6]$$

An anodic polarization is accompanied by an increase in the E potential of the electrode. A cathodic polarization leads to the electrode potential being decreased. Polarization, η, is defined by the overvoltage imposed with respect to the reversible equilibrium potential:

– anodic polarization: $\eta = E - E_{rev} > 0$;

– cathodic polarization: $\eta = E - E_{rev} < 0$.

The polarization curve of the reversible electrode is defined by the relationship between variables E and i of the electrochemical system. This curve describes an electrode's response to an imposed polarization (Figure 1.4).

In Figure 1.4, the current response (i) is represented by the black curve as a continuous line. It is important to note that the polarization curve of a system is defined within the strict framework of the steady-state condition (electric flux constant over time).

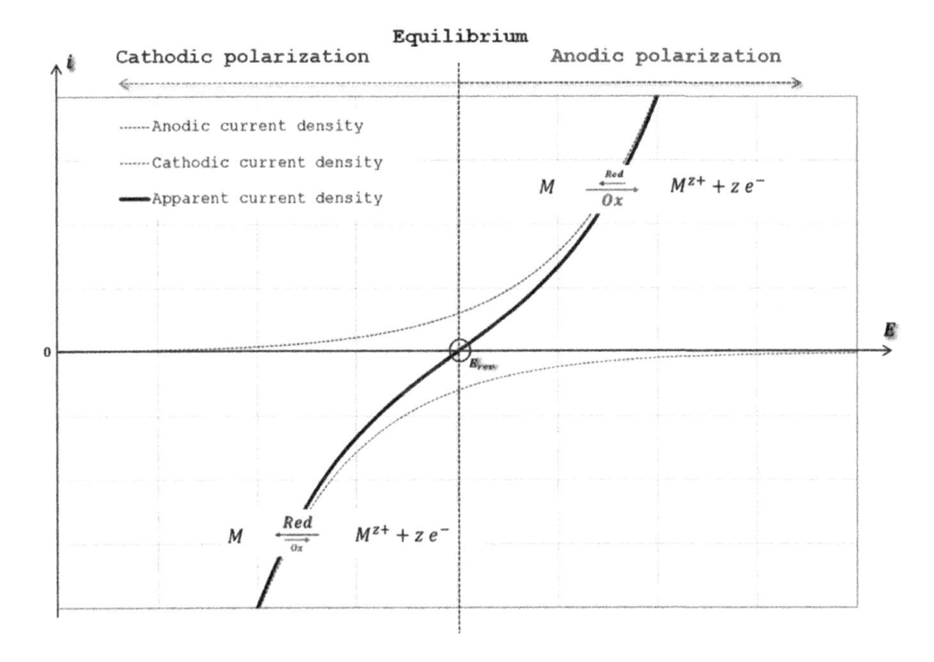

Figure 1.4. *Polarization curve of a reversible electrode (in black) and behaviors of anodic (red) and cathodic (blue) components. For a color version of the figure, see www.iste.co.uk/francois/corrosion.zip*

The response of a reversible electrode $i(E)$ is modeled by the Butler-Volmer (BV) equation (equation [1.7]), which links the apparent net current density (i) to the potential E of the electrode:

$$i = i_0 \left[\exp\left(\frac{E-E_{rev}}{\beta_a}\right) - \exp\left(-\frac{E-E_{rev}}{\beta_c}\right)\right] \quad [1.7]$$

This equation simply constitutes the algebraic sum of the anodic (ia) and cathodic (ic) currents referred to above, whose respective responses to a polarization are modeled by equations [1.8] and [1.9]:

$$i_a = i_0 \left[\exp\left(\frac{E-E_{rev}}{\beta_a}\right)\right] \quad [1.8]$$

and,

$$i_c = -i_0 \left[-\exp\left(\frac{E-E_{rev}}{\beta_c}\right)\right] \quad [1.9]$$

whereby:

– i_0 is the electrode's exchange current density (kinetic parameter),

– β_a and β_c are the electrode's anodic and cathodic Tafel coefficients, respectively.

The Butler-Volmer model of a reversible electrode is therefore characterized by two coupled state variables (i, E) and four parameters (E_{rev}, i_0, β_a and β_c).

In Figure 2.4, the curve shown as a red dotted line represents the response of the anodic reaction $i_a(E)$ and the curve shown as a blue dotted line reflects the response of the cathodic reaction $i_c(E)$. The system response ($i(E)$) is therefore composed of the algebraic sum of the anodic and cathodic responses. Thus, at equilibrium ($E = E_{rev}$), the anodic and cathodic currents compensate for one another and are equal, in absolute value, to the electrode's exchange current density: $i_a = -i_c = i_0$

Two noteworthy trends emerge, according to the type of action to which the electrode is subjected:

– anodic polarization amplifies anodic kinetics and reduces cathodic kinetics, which tends rapidly to 0 when $E \gg E_{rev}$ ($i_c \to 0$ and $i \to i_a$),

– cathodic polarization amplifies reduction kinetics and reduces anodic kinetics, which tends rapidly to 0 when $E << E_{rev}$ ($i_a \rightarrow 0$ and $i \rightarrow i_c$).

The Butler-Volmer equation is also frequently expressed as follows (equation [1.10]):

$$i = i_0 \left[\exp\left(\ln(10) \frac{E - E_{rev}}{b_a} \right) - \exp\left(-\ln(10) \frac{E - E_{rev}}{b_c} \right) \right] \quad [1.10]$$

whereby b_a and b_c represent the anodic and cathodic Tafel slopes, expressed in Volts per decade.

The Tafel slope concept is naturally illustrated when the polarization curve is represented in semi-logarithmic mode, in order to better grasp the exponential behavior (Figure 1.5), considering here the absolute value of the apparent net current density.

Figure 1.6 presents an example of a polarization curve drawn with the aid of arbitrary Butler-Volmer parameters and illustrates the respective effects of these parameters. It can be noted in particular that the coordinates of the point of intersection of the anodic and cathodic Tafel slopes correspond to the equilibrium potential and to the exchange current density.

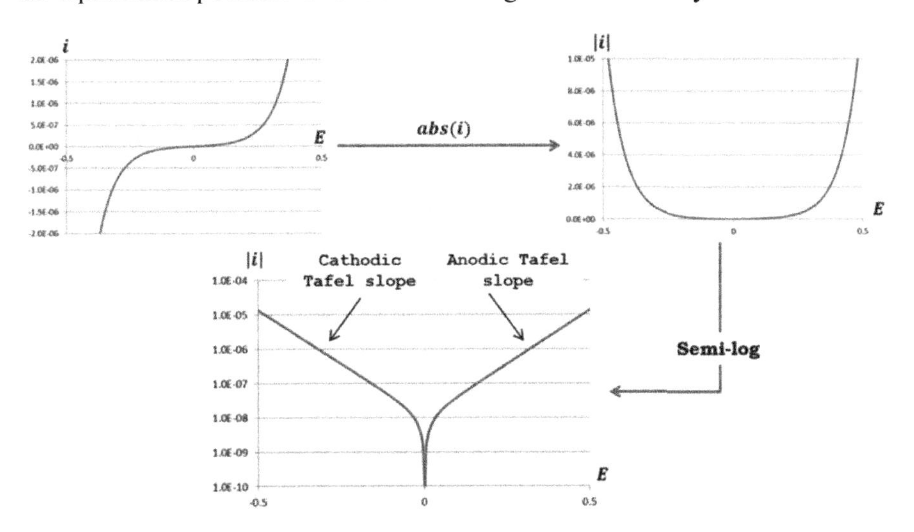

Figure 1.5. *Semi-log representation of the polarization curve of a reversible electrode. For a color version of the figure, see www.iste.co.uk/francois/corrosion.zip*

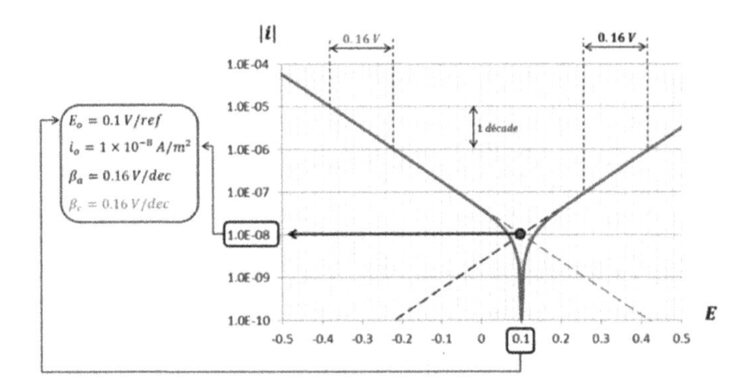

Figure 1.6. *Illustration of the effect of Butler-Volmer parameters on the polarization curve of a reversible electrode (semi-log mode). For a color version of the figure, see www.iste.co.uk/francois/corrosion.zip*

1.2.2. *Uniform corrosion system*

Uniform corrosion can be defined as the electrical coupling of two infinitely-close reversible electrodes. This definition implies that the spatial proximity of the two electrodes allows the ohmic effects due to electrolyte resistance to be overlooked. Where there is no electrical resistance between the anodic and cathodic sites, no potential gradient may exist in the space. As a consequence, the equilibrium potential of the coupled system is uniform. In particular, steel corrosion in concrete results from bringing two reversible electrodes, defined by the following equilibrium equations, into electrolytic contact:

– Electrode 1: $Fe \leftrightarrows Fe^{2+} + 2\, e^-$

– Electrode 2: $2\, OH^- \leftrightarrows \frac{1}{2} O_2 + H_2O + 2\, e^-$

Here the uniform character results from the fact that electrode 2 (O_2/OH^-) is present in dissolved form in the electrolyte (concrete) in contact with the steel. It can thus be assumed that electrode 2 is present across the entire surface of electrode 1. As illustrated in Figure 1.7, these two electrodes are characterized by different reversible potentials: $E_{rev2} > E_{rev1}$. It is to be noted that the table in Figure 1.7 presents the potentials measured under standard conditions. Under other conditions, the value of a reversible potential is defined by the Nernst law. Nevertheless, this observation does not call into question the theoretical developments below.

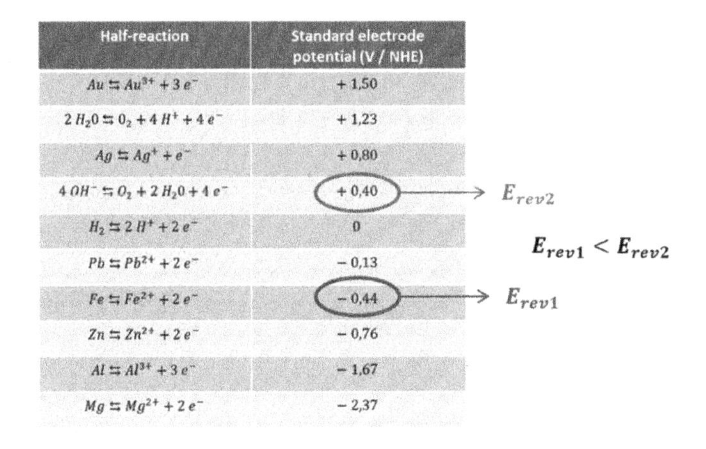

Half-reaction	Standard electrode potential (V / NHE)
$Au \leftrightharpoons Au^{3+} + 3\,e^-$	+ 1,50
$2\,H_2O \leftrightharpoons O_2 + 4\,H^+ + 4\,e^-$	+ 1,23
$Ag \leftrightharpoons Ag^+ + e^-$	+ 0,80
$4\,OH^- \leftrightharpoons O_2 + 2\,H_2O + 4\,e^-$	+ 0,40
$H_2 \leftrightharpoons 2\,H^+ + 2\,e^-$	0
$Pb \leftrightharpoons Pb^{2+} + 2\,e^-$	− 0,13
$Fe \leftrightharpoons Fe^{2+} + 2\,e^-$	− 0,44
$Zn \leftrightharpoons Zn^{2+} + 2\,e^-$	− 0,76
$Al \leftrightharpoons Al^{3+} + 3\,e^-$	− 1,67
$Mg \leftrightharpoons Mg^{2+} + 2\,e^-$	− 2,37

Figure 1.7. *Table of standard potentials. For a color version of the figure, see www.iste.co.uk/francois/corrosion.zip*

The coupling of these two electrodes allows the direct transfer of electrons from one to the other, resulting in a new state of equilibrium, which is now based on the system composed of the two coupled electrodes. In the present case, the hierarchy of the potentials of the two electrodes reveals that the anodic reaction favored by coupling corresponds to the dissolution of metallic iron to the benefit of a reduction in dioxygen. The Butler-Volmer model enables us to understand why this is so. It also enables precise definition of the state of equilibrium of the uniform corrosion system.

Figure 1.8 illustrates the establishment of the state of equilibrium of the system formed by the two reversible electrodes that are coupled and infinitely close. The red and blue curves represent the polarization curves of electrodes 1 and 2, respectively. The electrical coupling of the two electrodes results in their reciprocal polarization towards a common, free potential, referred to as the corrosion potential E_{corr}, and comprised between E_{rev1} and E_{rev2}. The free corrosion potential thus constitutes the system's equilibrium potential composed of the two infinitely-close, coupled electrodes. The hierarchy of the reversible potentials ($E_{rev1} < E_{rev2}$) implies that electrode 1 is anodically polarized (iron dissolution) and electrode 2 is cathodically polarized (reduction of dioxygen). *Corrosion appears here due to the fact that the anodic and cathodic reactions are no longer based on the oxidized and reduced forms of the same electrode.* Thus, the electrons freed as a result of iron dissolution are consumed to produce OH^- ions.

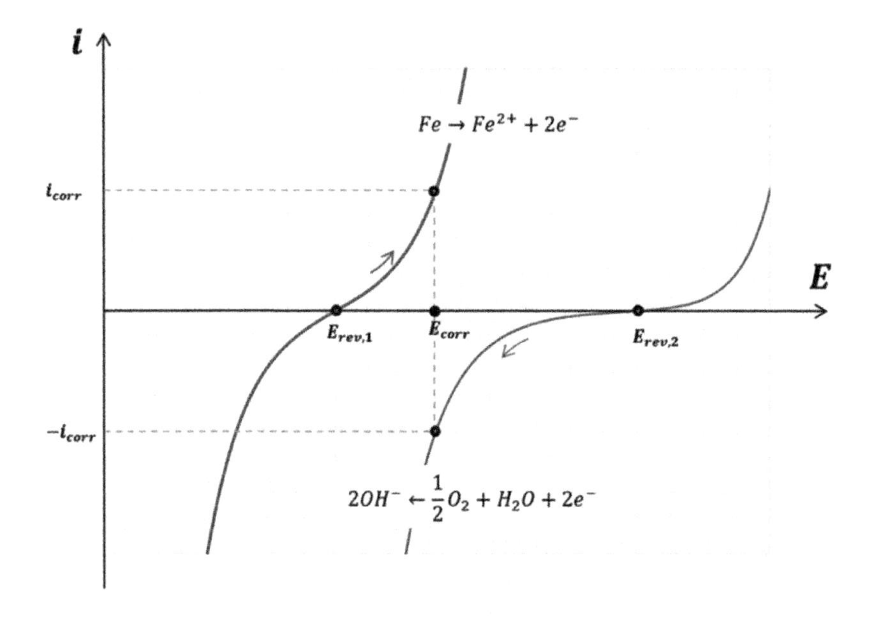

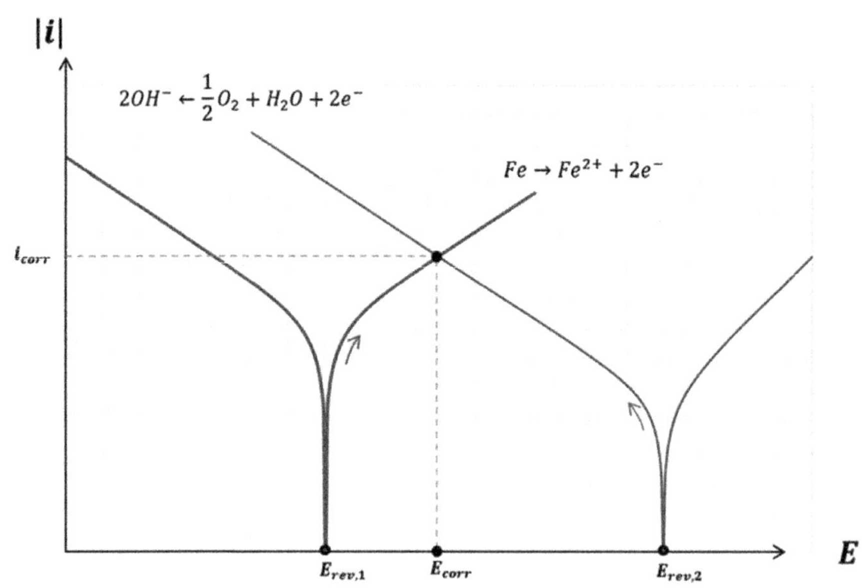

Figure 1.8. *Equilibrium of a uniform corrosion system – linear (top) and semi-log (bottom) representation modes. For a color version of the figure, see www.iste.co.uk/francois/corrosion.zip*

The corrosion potential resulting from coupling therefore represents an equilibrium potential that is both anodic and cathodic. Its value is dependent on the equilibrium between the anodic current produced by electrode 1 and the cathodic current produced by electrode 2. This condition defines a unique E_{corr} solution. At equilibrium, the anodic current density (opposite to the cathodic current density) defines the corrosion current density i_{corr}.

The state of equilibrium of the uniform corrosion system (E_{corr}) is entirely defined by the Butler-Volmer equations associated with each reversible electrode (equations [1.11] and [1.12]):

$$i_1 = i_{0,1}\left[\exp\left(\frac{E-E_{rev,1}}{\beta_{a,1}}\right) - \exp\left(-\frac{E-E_{rev,1}}{\beta_{c,1}}\right)\right] \quad [1.11]$$

and

$$i_2 = i_{0,2}\left[\exp\left(\frac{E-E_{rev,2}}{\beta_{a,2}}\right) - \exp\left(-\frac{E-E_{rev,2}}{\beta_{c,2}}\right)\right] \quad [1.12]$$

Assuming that the polarization to which each electrode is subjected is sufficiently strong, it is possible to overlook the respective cathodic and anodic terms of the equations relating to electrodes 1 and 2. Their behavioral laws can thus be summarized by (equation [1.13] and [1.14]):

$$i_1 = i_{0,1} \cdot \exp\left(\frac{E-E_{rev,1}}{\beta_{a,1}}\right) \quad [1.13]$$

and

$$i_2 = -\, i_{0,2} \cdot \exp\left(-\frac{E-E_{rev,2}}{\beta_{c,2}}\right) \quad [1.14]$$

At equilibrium, the apparent net current density (equation [1.15]) produced by the uniform corrosion system is zero:

$$i = i_1 + i_2 = i_{0,1} \cdot \exp\left(\frac{E_{corr}-E_{rev,1}}{\beta_{a,1}}\right) - \, i_{0,2} \cdot \exp\left(-\frac{E_{corr}-E_{rev,2}}{\beta_{c,2}}\right) = 0 \quad [1.15]$$

The latter equation contains a single unknown value (E_{corr}) and can therefore be solved in order to be determine the corrosion potential. The corrosion current density is then simply calculated using equation [1.16]:

$$i_{corr} = i_{0,1} \cdot \exp\left(\frac{E_{corr} - E_{rev,1}}{\beta_{a,1}}\right) \qquad [1.16]$$

At this stage, the equilibrium is precisely determined. Let us now introduce the Evans diagram, which constitutes an additional representation mode that also enables simple illustration of a uniform corrosion system composed by coupling two reversible electrodes (Figure 1.9). The Evans diagram can be likened to the semi-log representation defined above, except for two details:

– Axes i and E are permuted.

– Only the Tafel slopes of the different electrodes are represented.

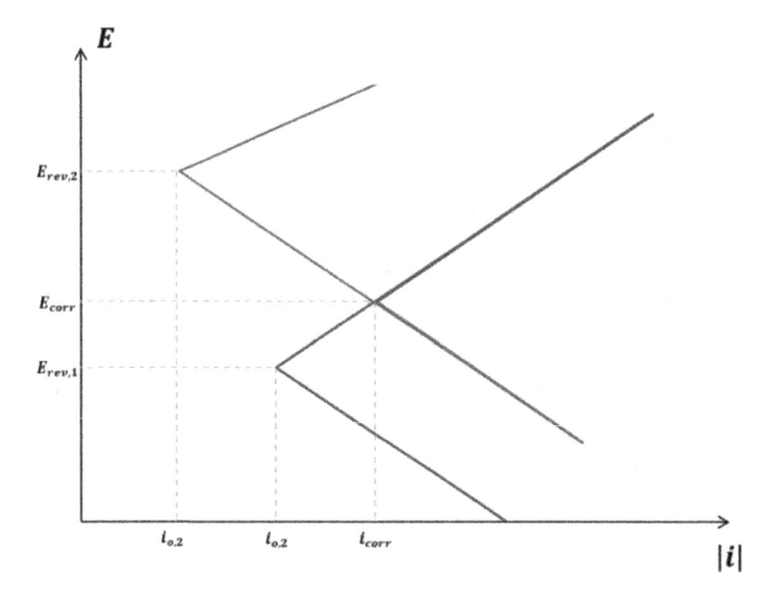

Figure 1.9. *Evans diagram of a uniform corrosion system. For a color version of the figure, see www.iste.co.uk/francois/corrosion.zip*

It can be seen from Figure 1.9 that the descriptive variables (E_{corr}, i_{corr}) of the uniform corrosion system at equilibrium can easily be deduced from the intersection of the anodic Tafel slope of electrode 1 and the cathodic Tafel slope of electrode 2.

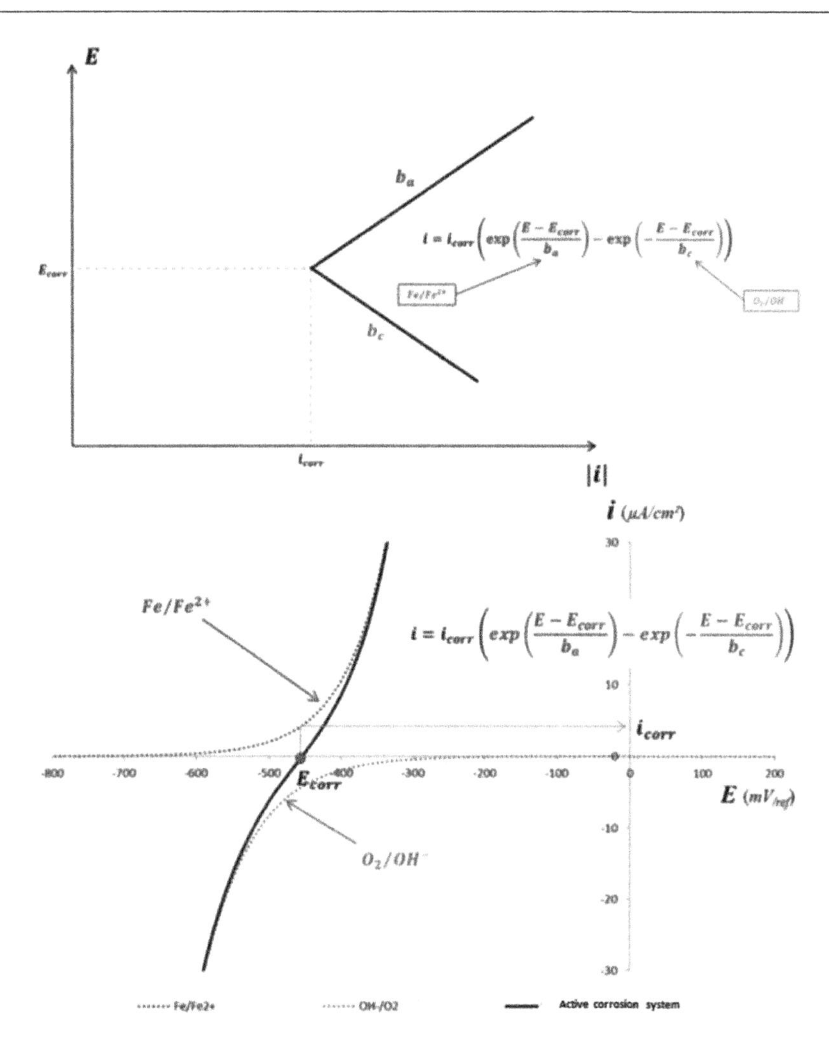

Figure 1.10. *Polarization curve of a uniform corrosion system. For a color version of the figure, see www.iste.co.uk/francois/corrosion.zip*

Upon observing the Evans diagram presented in Figure 1.9, it would appear that the response of a uniform corrosion system to an imposed polarization also observes Butler-Volmer's law (Figure 1.10). It is thus possible to define the BV equation of the uniform corrosion system, itself deriving from the BV equations of the reversible electrodes concerned (equation [1.17]). Here, this equation describes the response of the uniform corrosion system to an imposed polarization, with respect to its free potential E_{corr}. Subsequently, for the sake of simplicity, the Tafel constants will no longer be listed according to the associated reversible electrode and β_a will be used for $\beta_{a,1}$ and β_c for $\beta_{c,2}$.

$$i = i_{corr}\left[\exp\left(\frac{E-E_{corr}}{\beta_a}\right) - \exp\left(-\frac{E-E_{corr}}{\beta_c}\right)\right] \quad [1.17]$$

The polarization curve of the uniform corrosion system is illustrated in Figure 1.10 (curve drawn as a black continuous line). It therefore results from the sum of the BV equations, anodic (i_1) and cathodic (i_2) respectively, of reversible electrodes 1 and 2.

The linear polarization resistance R_p (equation [1.18]) of the uniform corrosion system is defined by the inverse slope of the polarization curve in E_{corr} (linear representation):

$$R_p = \left.\frac{\mathrm{d}E}{\mathrm{d}i}\right|_{E=E_{corr}} \quad [1.18]$$

The linear polarization resistance is expressed here in ohm.m^2. By linearizing the BV equation of the uniform corrosion system around its free potential E_{corr}, equation [1.19] is obtained:

$$i = i_{corr}\frac{\beta_a+\beta_c}{\beta_a\beta_c}(E - E_{corr}) \quad [1.19]$$

Using $\Delta E = E - E_{corr}$ to note the polarization to which the system is subjected, and $\Delta i = i - 0 = i$ for the associated apparent net current density, the corrosion current density may be expressed as follows (equation [1.20]):

$$i_{corr} = \frac{\beta_a\beta_c}{\beta_a+\beta_c}\frac{\Delta i}{\Delta E} \quad [1.20]$$

Where polarization is sufficiently low, the $\frac{\Delta E}{\Delta i}$ ratio is assimilated to the linear polarization resistance, R_p. By introducing $B = \frac{\beta_a\beta_c}{\beta_a+\beta_c}$, the Stern-Geary equation (equation [1.21]) is easily deduced from equation [1.20] [STE 57]:

$$i_{corr} = \frac{B}{R_p} \quad [1.21]$$

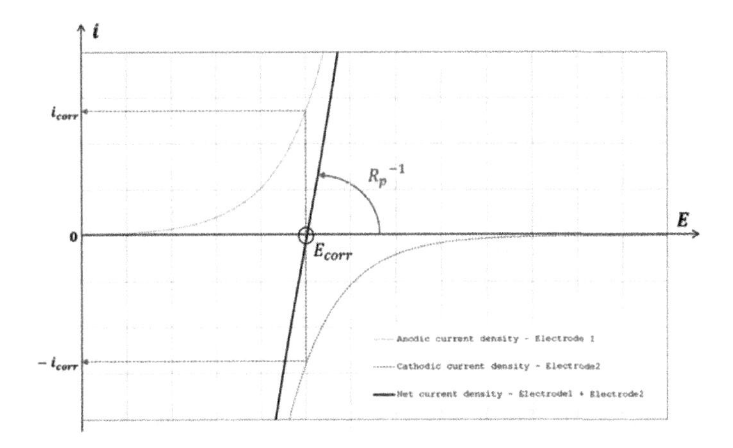

Figure 1.11. *Illustration of the concept of polarization resistance. For a color version of the figure, see www.iste.co.uk/francois/corrosion.zip*

This equation represents the theoretical basis of the kinetic measurement of a uniform corrosion system. The usual protocols for measuring R_p consist of imposing a polarization with respect to the corrosion potential and measuring the associated current, or vice versa. Once the polarization resistance has been measured, it is translated into corrosion current density via the Stern-Geary equation. This density is then converted into metallic mass loss using Faraday's law.

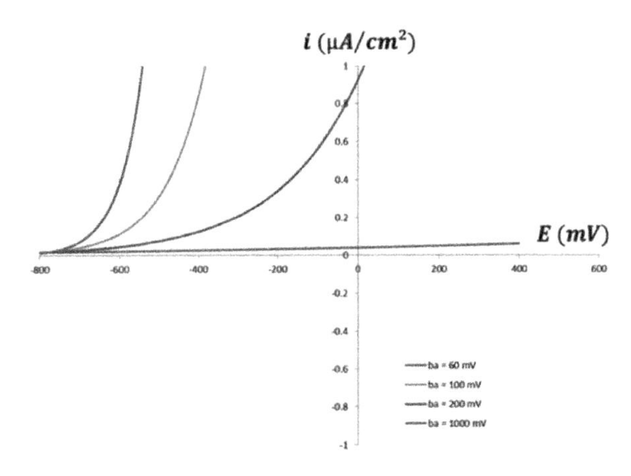

Figure 1.12. *Effect of the Tafel coefficient ba on the electrode's anodic curve Fe/Fe^{2+}. For a color version of the figure, see www.iste.co.uk/francois/corrosion.zip*

Butler-Volmer's equation makes it possible to model the responses of active and passive systems using the anodic Tafel coefficient β_a. Figure 1.12 provides a qualitative representation of the effect of the coefficient β_a on the anodic curve of the Fe/Fe^{2+} electrode. A low coefficient value reflects the behavior of active steel, presenting a very steep response curve (red curve). Inversely, a high anodic Tafel coefficient value enables the steepness to be significantly limited, such as to give the anodic response curve an almost-constant value.

Figure 1.13 presents two polarization curves obtained simply by changing the anodic Tafel coefficient value of the half-reaction Fe / Fe^{2+} (with all other system parameters set). The red and blue curves reflect the behavior in response to polarization of active and passive steel, respectively. It can be noted that an active system is characterized by a significantly lower corrosion potential than that of a passive system. Likewise, the ($E_{corr,i}$) curve slopes reveal a much lower polarization resistance in the case of an active system. Physically, this polarization resistance may be likened to an electric charge transfer resistance at the metal-electrolyte interface (transformation of an ionic current into electronic current). Thus, a high polarization resistance reveals a low charge transfer capability, and therefore low corrosion kinetics, and vice versa.

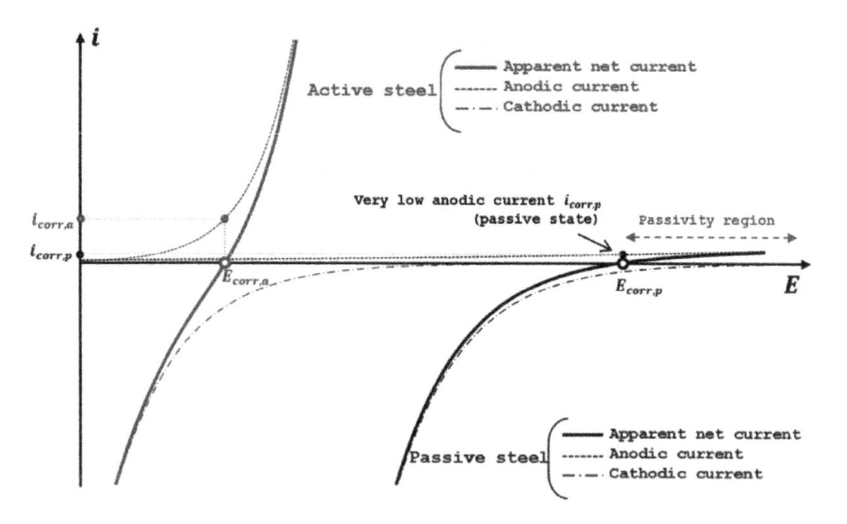

Figure 1.13. *Examples of polarization curves typical of uniform corrosion systems: active steel in red and passive steel in blue. For a color version of the figure, see www.iste.co.uk/francois/corrosion.zip*

1.2.3. *Localized (or galvanic) corrosion system*

Galvanic, or localized, corrosion, is defined as the electrical coupling of two spatially separate electrochemical systems. These systems may be made up of two reversible electrodes, two distinct corrosion systems (active and passive) or a reversible electrode coupled with a corrosion system. This definition implies that the distance between the two electrodes no longer allows the ohmic effects due to the electrical resistivity of the electrolyte (concrete) to be overlooked. As the anodic and cathodic sites are distant from one another, they cannot meet at a common potential, thus resulting in a significant difference between anodic and cathodic potentials, a potential gradient in the volume and a current, known as galvanic current or macro-cell current.

The state of equilibrium of a galvanic system is thus more difficult to define than in the case of a uniform corrosion system. Indeed, the galvanic system possesses an additional variable in that the anodic and cathodic potentials are different.

Let us, for pedagogical purposes, consider a galvanic system composed by the coupling of two uniform-corrosion systems. Let us return to the example of active and passive systems whose polarization curves were presented in Figure 1.13. The example adopted here may be likened to the galvanic coupling between a depassivated (active) area of steel in electrical contact with passive reinforcements. In order to simplify the illustrations, the following developments are based on a one-dimensional problem, i.e. the active sites are brought to a unique potential, E_a, and the passive sites are brought to a unique potential, E_p. In order to take account of the surface differences between anodic area and cathodic area, it becomes necessary to express the equilibrium of the anodic and cathodic currents in terms of intensity (I). A_a will be used for the active steel surface and A_p for the passive steel surface, respectively.

Like uniform corrosion, the electrical coupling of the two corrosion systems leads to mutual polarization of the active and passive sites (Figure 1.14). The active sites are anodically polarized and the passive sites are cathodically polarized. This results in an increase in the corrosion kinetics of the active steel by galvanic coupling and the cathodic protection of the passive steel.

The current responses of the active and passive sites to an imposed polarization are described by their BV equations (respectively: [1.22] and [1.23] and red and black curves in Figure 1.14):

$$I_a = A_a i_{corr,a}\left[\exp\left(\frac{E-E_{corr,a}}{\beta_{a,a}}\right)-\exp\left(-\frac{E-E_{corr,a}}{\beta_{c,a}}\right)\right] \quad [1.22]$$

$$I_p = A_p i_{corr,p}\left[\exp\left(\frac{E-E_{corr,p}}{\beta_{a,p}}\right)-\exp\left(-\frac{E-E_{corr,p}}{\beta_{c,p}}\right)\right] \quad [1.23]$$

In view of the polarization directions imposed by galvanic coupling, it is possible to overlook the cathodic component of active steel and the anodic component of passive steel, which leads to equations [1.24] and [1.25]:

$$I_a = A_a\, i_{corr,a}\, \exp\left(\frac{E-E_{corr,a}}{\beta_{a,a}}\right) \quad [1.24]$$

$$I_p = -A_p\, i_{corr,p}\, \exp\left(-\frac{E-E_{corr,p}}{\beta_{c,p}}\right) \quad [1.25]$$

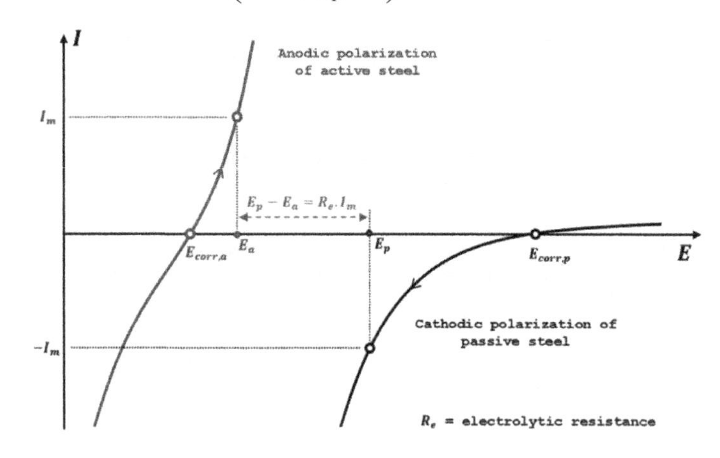

Figure 1.14. *Equilibrium of a 1D localized corrosion system. For a color version of the figure, see www.iste.co.uk/francois/corrosion.zip*

At the equilibrium of the galvanic system, the currents associated with the active and passive sites compensate for one another in order to preserve the overall electroneutrality (equation [1.26]):

$$I_a(E_a) + I_p(E_p) = 0 \quad [1.26]$$

That is:

$$A_a\, i_{corr,a} \exp\left(\frac{E_a - E_{corr,a}}{\beta_{a,a}}\right) - A_p\, i_{corr,p} \exp\left(-\frac{E_p - E_{corr,p}}{\beta_{c,p}}\right) = 0 \quad [1.27]$$

Unlike in the case of uniform corrosion, equation [1.27], defining the compensation of anodic and cathodic currents, does not enable the state of equilibrium of the galvanic system to be determined (one equation for two unknowns, E_a and E_p). It is therefore of indeterminate form, admitting an infinity of solutions (E_a, E_p).

In order to express the equilibrium of a galvanic system it is therefore necessary to turn to a complementary equation, on this occasion Ohm's law. Indeed, the equilibrium of such a system naturally depends on its electrochemical components as well as on the electrical resistivity of the electrolyte and of the system's spatial structure.

At equilibrium, a galvanic current, I_m, (equation [1.28]) flows in the volume from the anodic sites to the cathodic sites:

$$I_m = A_a\, i_{corr,a} \exp\left(\frac{E_a - E_{corr,a}}{\beta_{a,a}}\right) \quad [1.28]$$

This current encounters an ohmic resistance, R_e, due to the resistivity of the electrolyte and Ohm's law is expressed by equation [1.29] (1D case):

$$E_p - E_a = R_e\, I_m = R_e\, A_a\, i_{corr,a} \exp\left(\frac{E_a - E_{corr,a}}{\beta_{a,a}}\right) \quad [1.29]$$

The state of equilibrium (E_a, E_p) of a 1D galvanic system is therefore thoroughly derived from the following system of non-linear algebraic equations (equation [1.30]):

$$\begin{cases} A_a\, i_{corr,a} \exp\left(\frac{E_a - E_{corr,a}}{\beta_{a,a}}\right) - A_p\, i_{corr,p} \exp\left(-\frac{E_p - E_{corr,p}}{\beta_{c,p}}\right) = 0 \\ E_p - E_a = R_e\, A_a\, i_{corr,a} \exp\left(\frac{E_a - E_{corr,a}}{\beta_{a,a}}\right) \end{cases} \quad [1.30]$$

For a 3D galvanic system, the equilibrium is determined in an identical manner. The above equation system is simply reformulated as a differential boundary value problem, considering anodic and cathodic potential fields instead of scalar values, the local Ohm's law and the electric charge conservation equation.

1.3. Specific features of steel corrosion in concrete

1.3.1. *Passivation of steel in concrete*

The reinforcements used for reinforced concrete are plain or deformed steel bars, in that ribs or reliefs on the surface improve the quality of the bond with concrete. Plain bars tend to be used more as the assembly steel for the reinforcement cage, whereas deformed reinforcements are positioned in the tensile concrete areas. Commonly these reinforcements are made of uncoated, non-alloyed carbon steel. Carbon alloy reinforcements exist nevertheless, as well as reinforcements coated in another metal, such as galvanized steel with a zinc layer.

When leaving the factory, the reinforcements are covered with a layer of mill scale. Whilst stored prior to placement, they oxidize according to a corrosion mode known as *atmospheric* and become covered in a uniform rust film. This highly porous layer formed on the surface does not, however, jeopardize the casting of the concrete and even appears to assist the formation of a passive layer presenting a greater resistance to depassivation [CHA 18].

Once the reinforcement cage has been put in place inside the formwork and the concrete has been poured, the initial products of atmospheric corrosion transform and passivation occurs on the steel as a highly-intense corrosion process starts. A thin layer of corrosion products then rapidly forms at the steel-concrete interface. This layer of oxides, which is highly dense and only slightly soluble, forms a passivation layer at the steel's surface. As the iron Pourbaix diagram indicates, the formation and thermodynamic stability of the passivation layer can be explained by the highly alkaline nature of the concrete interstitial solution ($pH > 13$). As soon as the film is generated, it forms a physical barrier between the steel and the cover concrete. The corrosion kinetics is then reduced to a negligible level with regard to the standard service-life of civil engineering structures. Nevertheless, passivation must not be mistaken for immunity. The passive state corresponds to a corrosion state presenting negligible kinetics in the vast majority of cases of reinforced-concrete constructions. In certain specific cases of structures with a long service life, corrosion in the passive state nevertheless needs to be taken into account in durability predictions.

Generally speaking, the theoretical process for the formation of a passive film leads to the creation of a double layer made up of metal oxides directly in contact with the metal (the inner layer), and hydroxides in contact with the electrolyte (outer layer). In practice, experimental observations tend to confirm that metal oxides alone form the inner layer, but are more nuanced with regard to the composition of the outer layer. Thus, the steel passivation film in concrete is mainly formed by magnetite (Fe_3O_4) in the immediate vicinity of steel. Moving away from the steel to the outside of the passive layer, it was possible to observe the two oxidized forms: hematite ($\alpha - Fe_2O_3$) formed locally, $FeOOH$-type hydroxides and maghemite ($\gamma - Fe_2O_3$) [AND 01]. Studies show that the thickness of the passive layer varies from 1 to 20 nm [HUE 05].

This passivation phenomenon is one of the keys to the success encountered by reinforced-concrete composite in the history of construction. Steel makes up for concrete's lack of tensile strength and concrete provides steel with high chemical resistance owing to the alkaline environment making it up. Along with the reasonable financial cost, this reciprocity explains why reinforced concrete has ranked first among modern construction materials for several decades.

The sustainability of this composite may, however, be affected by two independent physical mechanisms capable of leading to the dissolution of the passivation film and the initiation of active corrosion of steel reinforcements. These two mechanisms result when aggressive agents penetrate the concrete porous space: atmospheric carbon dioxide (CO_2) and chloride ions (Cl^-).

1.3.2. *Corrosion initiation*

Two distinct, independent causes lead to corrosion initiation through local depassivation of concrete steel reinforcements:

– Concrete contamination by chloride ions.

– Concrete carbonation by atmospheric CO_2.

1.3.2.1. *Concrete contamination by chlorides*

The exogenous chlorides that bring about corrosion problems in reinforced-concrete structures essentially originate from two sources [NEV 95, MON 03]:

– A marine environment, where concrete may be contaminated, either through direct contact with sea water, or through sea spray.

– De-icing salts used in harsh winter conditions.

Endogenous chlorides result from the components of concrete: cement, water, additives and aggregates. These chlorides present at hardening are then for the most part fixed by hydration reactions, hence Standard NF EN 206 [STA 14] is able to authorize the presence of a maximum rate of chlorides in cement, according to the structure type. By contrast, the use of chlorinated additives, sea water or aggregates polluted by chlorides is prohibited.

Local depassivation is likely to occur in steel when the chlorides enter into direct contact with the passive reinforcement film. For this to happen, they must cross the layer of cover concrete to reach the reinforcement layout. Part of the chlorides diffused into the concrete is fixed by the cement matrix and does not contribute to depassivation. Only free chlorides, present in the interstitial solution, are theoretically available to activate corrosion. When the free-chloride concentration in the vicinity of the reinforcement reaches a critical value, the passive film is damaged locally and corrosion is then initiated. The role of chlorides on the passive layer of steel reinforcements covered by concrete remains uncertain: three main mechanisms with the potential to act together are discussed [STE 95]: (1) mechanism of adsorption (physical and chemical), with catalytic effect, of chlorides to dissolve oxides, (2) mechanism of penetration of chlorides via passive-layer defects, (3) mechanism of passive-layer failure through electrostriction due to a decrease in surface tension resulting from chloride adsorption.

Much research has been dedicated to identifying a chloride critical threshold that would *initiate* reinforcement corrosion, but no consensus has been found as yet [IZQ 04, ANG 17a, ANG 17b]. It would appear that the very concept of a given amount of chlorides initiating corrosion (whether deterministic or probabilistic) is not appropriate as there are other parameters that play a very significant role. This is particularly applicable to local conditions at the steel-concrete interface and to the composition and nature of the cement matrix. The term "interface" includes the parameters and defects specific to steel (inclusions, metallurgical composition, mill-scale layer and pre-oxidation layer due to worksite storage), those specific to concrete (composition of the cement paste, aggregates, reinforcing-steel spacers, cracks and porosity) and those specific to the interface (transition zone (ITZ), gaps, bleeding area, porosity and humidity gradient) [ANG 17b].

On these very localized, initial corrosion sites, the Cl^- ions react with the Fe^{2+} ions to form iron chloride, $FeCl_2$. The iron(II) hydroxide $Fe(OH)_2$ created by the consumption of hydroxyls from the interstitial solution ultimately results in the formation of green rust (Figure 1.15). Depending on the chloride content, four main products may form [GEN 97]: goethite, maghemite, lepidocrocite (γ-FeOOH) and akaganeite (β-FeOOH). These relatively fluid corrosion products move through the porous space until possibly appearing at the concrete surface.

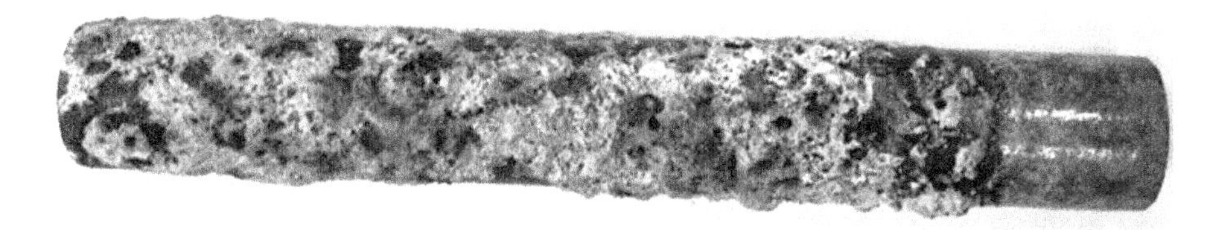

Figure 1.15. *Green rust. For a color version of the figure, see www.iste.co.uk/francois/corrosion.zip*

As mentioned above, chlorides induce localized corrosion. They depassivate the reinforcement layout locally, thus generating galvanic coupling between the new sites of active steel and the surrounding sites of passive steel (Figure 1.16). Once the galvanic cell has been initiated, the phenomenon is self-maintained as the electrical field created tends to concentrate chlorides in the active areas. The localized anodic sites exchange electrical current with the passive steel areas, where the cathodic reaction occurs. The current streamlines shown in Figure 1.16 represent the galvanic corrosion current. The current is ionic in concrete and electronic in the steel-reinforcement layout (Figure 1.16).

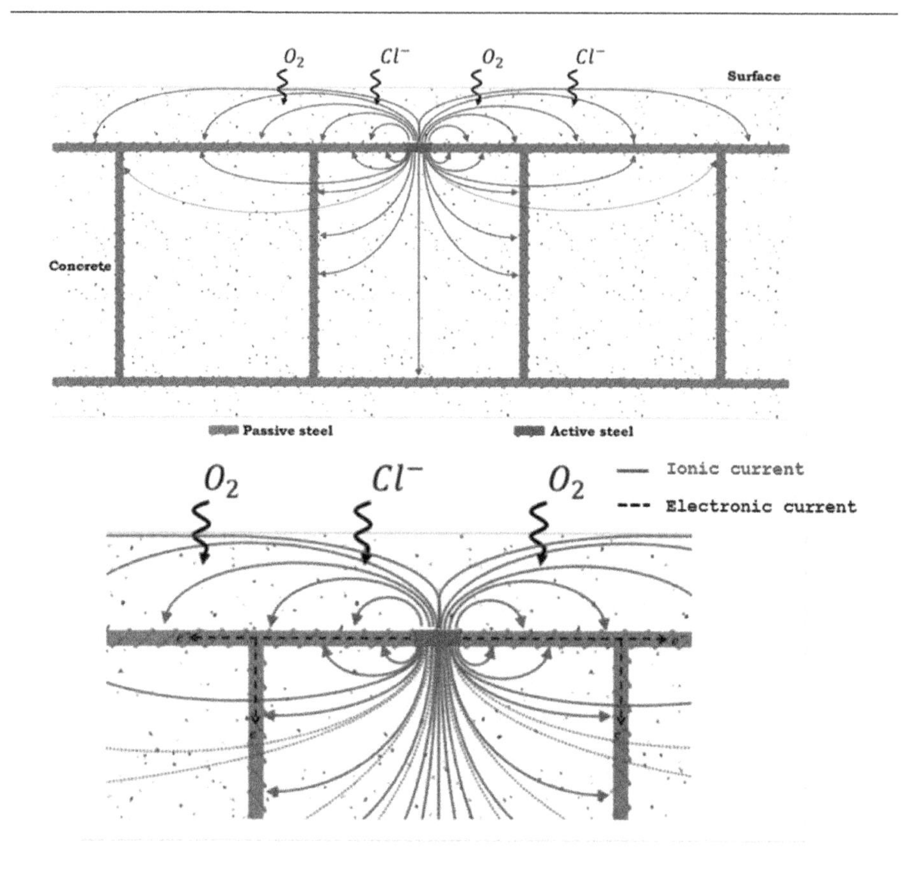

Figure 1.16. *Illustration of localized corrosion induced by chlorides. For a color version of the figure, see www.iste.co.uk/francois/corrosion.zip*

1.3.2.2. *Concrete carbonation*

Carbonation results from the penetration of the atmospheric CO_2 within the concrete pore space. Carbon dioxide dissolves in the interstitial solution and forms a weak acid (carbonic acid). The latter then reacts with the hydrates of the cement paste (Portlandite $Ca(OH)_2$ and C-S-H $CaO.SiO_2.nH_2O$) to form calcite ($CaCO_3$), which partially fills the porous volume and generally improves the concrete's mechanical properties. However, these reactions consume hydroxyl ions and are thus accompanied by a reduction in pH to a value comprised between 8 and 9. From a thermodynamic perspective, the Pourbaix iron diagram indicates that such a range of pH leads to the destabilization and ultimately dissolution of the passivation film.

The corrosion products formed in the case of carbonation reveal two areas at the steel-concrete interface [CHI 05]. A first, dense layer of corrosion products is formed upon contact with steel, mainly composed of goethite (α-FeOOH) but also containing other oxides, such as maghemite and/or magnetite. There then follows a second, mixed layer containing corrosion products that have diffused in the cement paste.

The vast majority of the scientific community holds the position that carbonation induces uniform corrosion. This view is likely based on the visible consequences of this mode of corrosion, which generates apparently uniform damage (of the concrete and steel) at the first reinforcement layer. Another argument often presented to justify the assumption of uniform corrosion in the case of carbonation, focuses on the strong electrical resistivity of carbonated concrete, which would prevent galvanic coupling. However, the latter assumption overlooks a phenomenon that is indisputable in reality. The carbonation front progression from the surface to the core of the structure is an extremely slow phenomenon. As a result, concrete is hardly ever fully carbonated within an actual structure, with only the first few centimeters of the cover concrete likely to be. Moreover, it would take several years for the carbonation front progression to cover the distance of the reinforcement diameter. Under such conditions, only part of the circumference of the bar is depassivated at the moment when the carbonation front reaches it and it would take several years to depassivate the entire bar. It is important to note, however, that in the presence of interface defects relating, for example, to the casting of fresh concrete (as discussed in Chapter 2), the interface can carbonate entirely as soon as the carbonation front reaches the reinforcement. Lastly, a recent experimental study on specimens of partially-carbonated reinforced concrete unequivocally negates the supposed impact of carbonated-concrete resistivity on the development of galvanic currents between depassivated areas and passive areas [SOH 15]. Indeed, significant galvanic currents have been measured between an active bar dipped into the carbonated part and a passive bar located within the sound area, with these strong galvanic currents crossing a 2-cm layer of carbonated concrete.

Under such conditions, the assumption of uniform corrosion in the electrochemical sense does not match the effects of carbonation. In the case of a partially carbonated structure, galvanic exchanges are possible between active areas and passive areas and the assumption of localized (or galvanic) corrosion is more appropriate since the anodic and cathodic sites are clearly dissociated at structure scale (Figure 1.17).

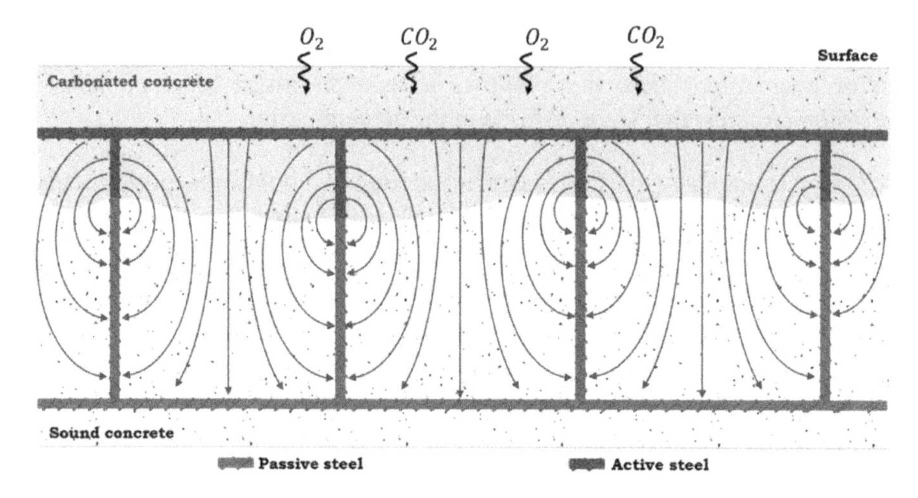

Figure 1.17. *Illustration of localized corrosion induced by concrete carbonation. For a color version of the figure, see www.iste.co.uk/francois/corrosion.zip*

Thus, whatever the cause of the phenomenon (chlorides or carbonation), it appears that steel corrosion in concrete is localized in nature in the vast majority of cases. The essential difference between these two fundamental causes of corrosion lies in the extent of the anodic surface area, which is very localized in the case of chlorides and more extended in the case of carbonation.

1.3.2.3. *3D physical problem and modeling*

As mentioned above, steel corrosion in reinforced concrete is localized (or galvanic) in nature in the vast majority of cases, regardless of the cause of initiation. Every case of a corroded structure is unique and does not allow for any geometric simplification. Galvanic equilibrium between an anodic site initiated locally and the passive surface mobilized within its environment for the cathodic reaction represents a three-dimensional physical problem.

In face of belated awareness of this reality, the modeling and numerical simulation of corrosion systems in reinforced concrete and the associated measurement techniques have formed the subject of fairly intense research works for only the last 20 years or so [GUL 99, KRA 01, WAR 06, NAS 10, WAR 10, CLÉ 12, SOH 15, LAU 16, MAR 16, SAS 16]. As a first approximation, the solution to a given problem, in other words a certain spatial configuration of active and passive sites within the concrete volume, is governed by local Ohm's law (equation [1.31]) and equation [1.32] expressing electric charge conservation within the volume:

$$\boldsymbol{i} = \frac{1}{\rho}\boldsymbol{E} = -\frac{1}{\rho}\nabla V \qquad [1.31]$$

$$\nabla.\boldsymbol{i} = 0 \qquad [1.32]$$

whereby:

– $\boldsymbol{i}$ is the current density vector (A/m^2);

– $\boldsymbol{E}$ is the electrical field vector (V/m);

– V is the potential field (V);

– ρ is the concrete electrical resistivity (Ohm.m).

The active and passive sites are modeled as relative boundary conditions at the steel-concrete interfaces by the Butler-Volmer equation. The Butler-Volmer parameters may be adapted locally in order to appropriately describe the active or passive state. Figures 1.18 and 1.19 illustrate a straightforward example of a numerical simulation using the finite-element method [LAU 16].

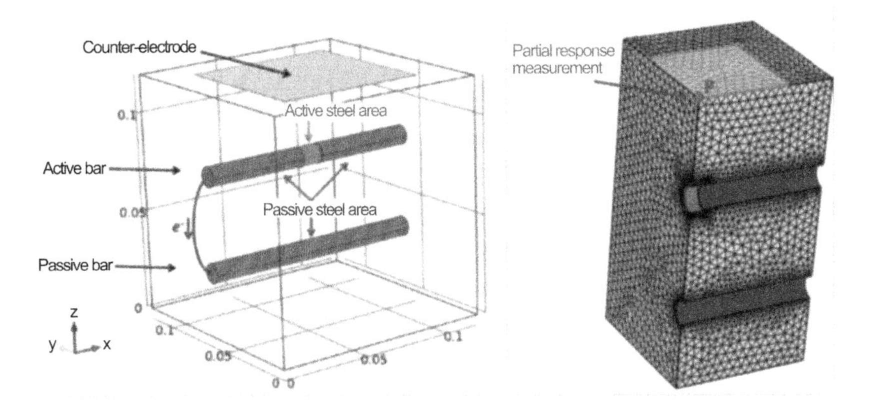

Figure 1.18. *Simulation of localized corrosion system: geometric model [LAU 16]. For a color version of the figure, see www.iste.co.uk/francois/corrosion.zip*

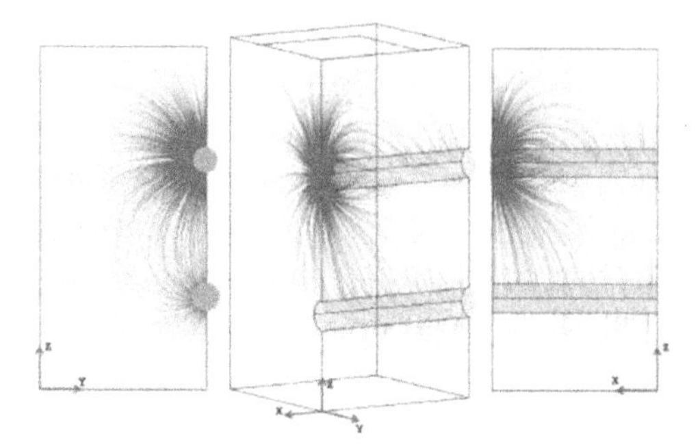

Figure 1.19. *Simulation of localized corrosion system: corrosion current streamlines [LAU 16]. For a color version of the figure, see www.iste.co.uk/francois/corrosion.zip*

1.3.2.4. *Corrosion kinetics control mechanisms*

The corrosion process depends on the anodic reactions, cathodic reactions and ionic and electronic transports between the anodic and cathodic sites. The limiting factors of corrosion kinetics are numerous: the anodic process, the cathodic process with, in particular, dioxygen availability, the anode/cathode ratio and the ohmic resistance between anode and cathode. In view of these limiting factors, the impact of electronic transportation in reinforcements is negligible.

1.3.2.4.1. Mobilizable passive surface, anode/cathode surface-area ratio and steel-reinforcement density

The establishment of the equilibrium of a galvanic corrosion system is based on the intensities of the anodic current produced by the active sites and the cathodic current produced by the passive sites being equal. For an anodic site of a given size, the intensity of the current produced depends on the extent of the passive surface mobilized for the cathodic reaction. Thus, the larger the mobilized passive surface, the stronger the corrosion kinetics (anodic current). The *mobilized passive surface* thus constitutes the cathode of the corrosion system. Generally, the impact of the anode/cathode (A/C) surface ratio on corrosion kinetics is referred to; a low A/C ratio generates rapid corrosion. Inversely, a high A/C ratio is accompanied by slower corrosion.

In reinforced concrete, the mobilizable passive surface for the cathodic reaction is conditioned by certain characteristics that thus constitute the fundamental influencing factors of corrosion kinetics:

– dlectrical resistivity and water content of concrete;

– dioxygen availability;

– steel-reinforcement density of the structural element.

The effects of electrical resistivity and of dioxygen availability on the mobilizable passive surface and therefore on corrosion kinetics are described in the following two sections of this chapter. The element's steel-reinforcement density exerts a clear influence on the overall electrochemical exchange surface and therefore on the A/C ratio. A corrosion site initiated on a steel-reinforcement layout with a high volumetric density will be subject to greater anodic polarization owing to the large surface area of the surrounding passive steel.

1.3.2.4.2. Electrical resistivity and water content of concrete

Section 1.2.3 highlighted the impact of electrical resistivity on the equilibrium of a galvanic system and on the associated corrosion current. In concrete, variations in electrical resistivity are linked primarily to fluctuations in volumetric water content [SAL 96]. For a 1D-model corrosion system, Figure 1.20 illustrates this effect for two resistivity values using the Evans diagram: $\rho_1 > \rho_2$. When the resistivity increases, the mutual polarization of the active and passive sites is weaker. As a consequence, the current supplied by the active site decreases ($I_{m,1} < I_{m,2}$) and the potential gradient in the volume increases ($E_{p,1} - E_{a,1} > E_{p,2} - E_{a,2}$).

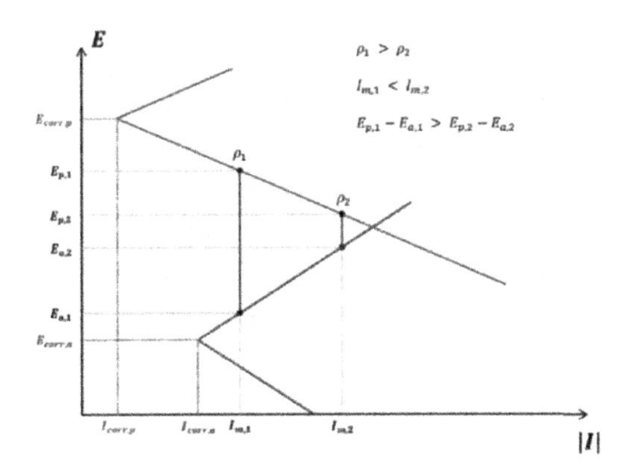

Figure 1.20. *Effect of electrical resistivity of concrete on galvanic coupling. For a color version of the figure, see www.iste.co.uk/francois/corrosion.zip*

This effect of resistivity on the galvanic coupling between active and passive sites, illustrated here using an ideal, one-dimensional example, is also valid from a qualitative point of view for actual, three-dimensional corrosion systems. In order to understand in simple terms the impact of resistivity on the galvanic kinetics of a 3D system, it would be trivial to reason using the concept of mobilizable passive surface defined above. Indeed, the lower the resistivity, the less resistance the concrete offers to galvanic-current transport and the larger the mobilized passive surface area around an active site. In other words, low resistivity solicits the cathodic reaction over a significant distance around the active site, thus producing a strong corrosion current. Inversely, high resistivity mobilizes a smaller passive surface area and is therefore accompanied by low corrosion kinetics.

Some authors provide experimental evidence of a "linear" relationship as a Log-Log representation between concrete electrical resistivity and corrosion kinetics (Figure 1.21).

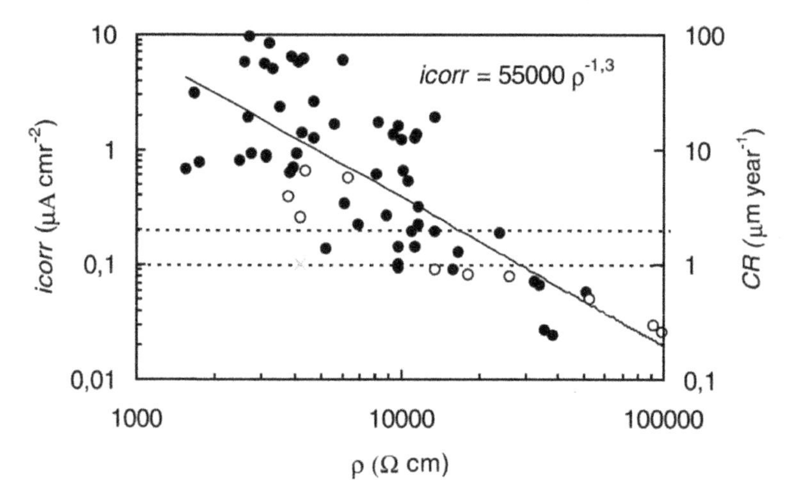

Figure 1.21. *Example of an experimental relationship between electrical resistivity and corrosion current [MOR 02]*

In order to test the relevance, from a theoretical point of view, of this relationship, the graph in Figure 1.22 summarizes the results of a series of numerical simulations conducted on four models of galvanic systems that differ in terms of anodic area size. For each system, three simulations are performed, with the concrete electrical resistivity set to the following values:

100, 200 and 400 Ω.m. The galvanic corrosion kinetics is reflected here by the intensity of the current supplied by the anodic area. There are two noteworthy results worth remarking on:

– for an anodic site of a given size, the numerical simulation based on the physical models described above does indeed confirm the theoretical existence of this linear dependency between the log of the corrosion current produced and the electrical-resistivity log;

– the size of the anodic site has an impact on the corrosion current via the A/C ratio, but only has a low impact on the slope of the Log-Log linear trend.

Based on this second remark, the apparent scatter of the experimental data presented in [MOR 02] can thus be explained by the fact that the sizes of the anodic sites of the different experimental specimens were neither controlled nor known. In Figure 1.22, relating to the above-mentioned numerical experiment, if only the points are plotted without labeling on the size of the anodic site to which they correspond, a graph very similar to that of Figure 1.21 is obtained.

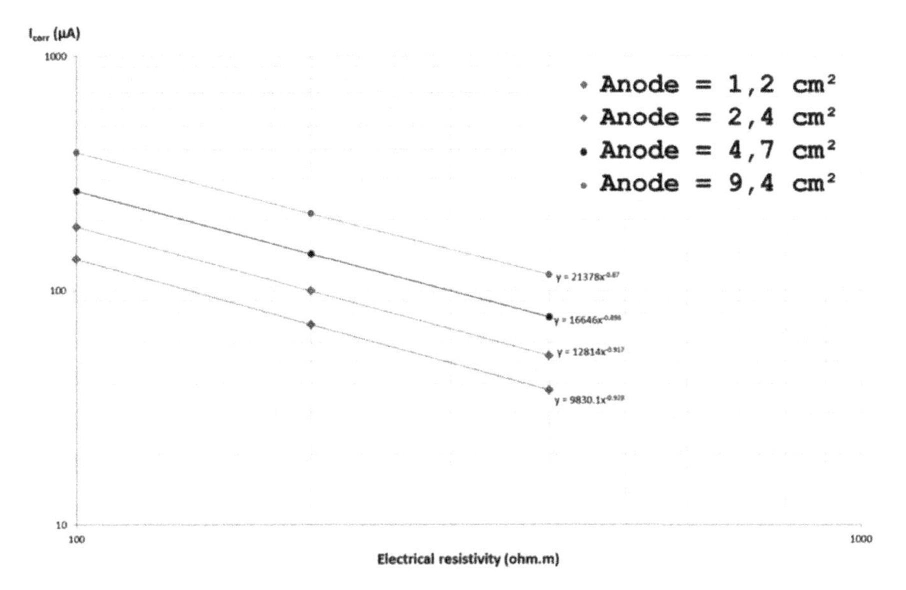

Figure 1.22. *Numerical simulation of the effects of electrical resistivity and of the size of the anodic site on the galvanic corrosion current. For a color version of the figure, see www.iste.co.uk/francois/corrosion.zip*

1.3.2.4.3. Dioxygen availability at the cathode

Cathodic reaction naturally plays an important role in the establishment of the equilibrium of a corrosion system (uniform or localized). Physically, the cathodic current corresponds to the amount of dioxygen reduced by time unit (or alternatively of hydroxyl ions formed by time unit), at the steel-concrete interface.

Previous discussions regarding the establishment of the equilibrium of a corrosion system implicitly assumed there to be an unlimited amount of dioxygen available for the cathodic reaction. In practice, the dissolved dioxygen is not always available in large amounts in the interstitial solution in contact with the steel. This availability is conditioned in particular by the dioxygen's ability to diffuse through the cement matrix.

In the case of a limited amount of dioxygen, the cathodic curve loses its exponential nature beyond a certain level of polarization and saturation of the cathodic current is observed. The value of the cathodic current saturation is defined as the cathodic limiting current i_{lim}.

Figure 1.23 presents a typical polarization curve for a passive steel (black curve), resulting from the algebraic sum of the anodic curve (red) and the cathodic curve (blue). As set out above, the Butler-Volmer equations of the anodic and cathodic curves are expressed with equations [1.33], [1.34] and [1.35]:

$$i(E) = i_a(E) + i_c(E) \qquad [1.33]$$

with:

$$i_a(E) = i_{corr} \exp\left(\frac{E-E_{corr}}{\beta_a}\right) \qquad [1.34]$$

$$i_c(E) = -\, i_{corr} \exp\left(-\frac{E-E_{corr}}{\beta_c}\right) \qquad [1.35]$$

In the case of a limited dioxygen supply, the cathodic component of Butler-Volmer's equation [1.36] for the corrosion system may be modified (equation [1.37]) to take account of the existence of the cathodic limiting current i_{lim}, conditioned by the dioxygen diffusion kinetics as far as the reinforcement layout:

$$i(E) = i_a(E) + i'_c(E) \quad [1.36]$$

whereby:

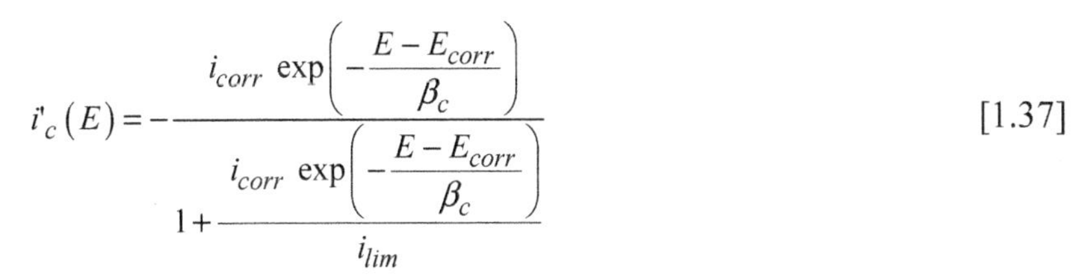

$$i'_c(E) = -\frac{i_{corr}\exp\left(-\frac{E - E_{corr}}{\beta_c}\right)}{1 + \frac{i_{corr}\exp\left(-\frac{E - E_{corr}}{\beta_c}\right)}{i_{lim}}} \quad [1.37]$$

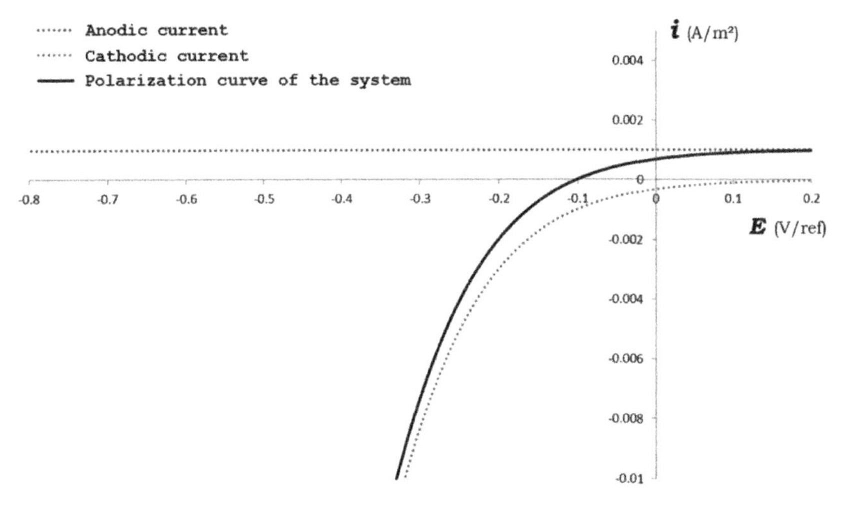

Figure 1.23. *Typical polarization curve for passive steel with unlimited amount of dioxygen. For a color version of the figure, see www.iste.co.uk/francois/corrosion.zip*

Figure 1.24 presents the typical value of a polarization curve for a passive steel with limited dioxygen availability.

Figure 1.25 presents the impact of parameter i_{lim} on the polarization curve of a passive steel. It can be noted here that the equilibrium potential of the passive system (E_{corr}) decreases with the reduction in dioxygen availability.

In actual structures, the existence of dioxygen concentration gradients thus generates potential gradients in the absence of corrosion. This phenomenon is typically encountered in partially-immersed structures for which the dissolved-dioxygen concentration is low in the immersed part and high in the emerged part.

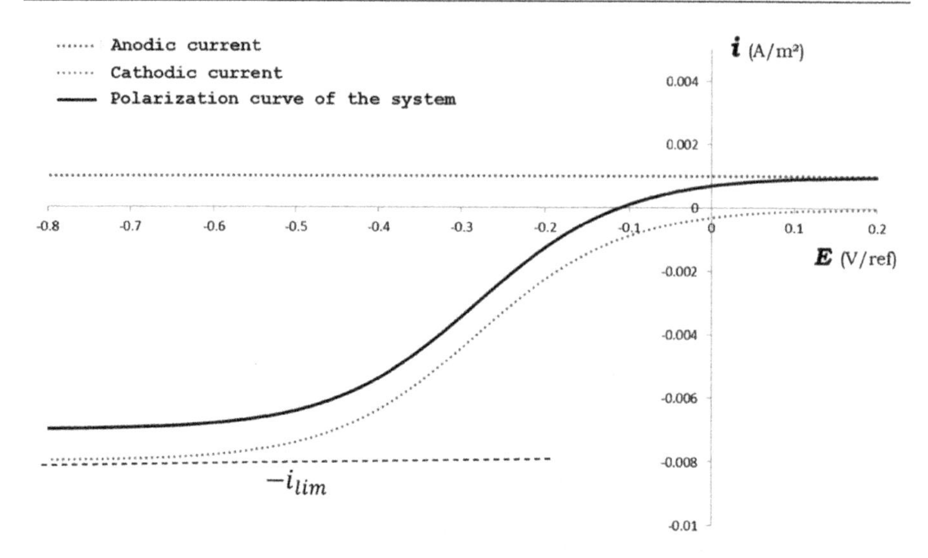

Figure 1.24. *Typical polarization curve for passive steel with limited amount of dioxygen. For a color version of the figure, see www.iste.co.uk/francois/corrosion.zip*

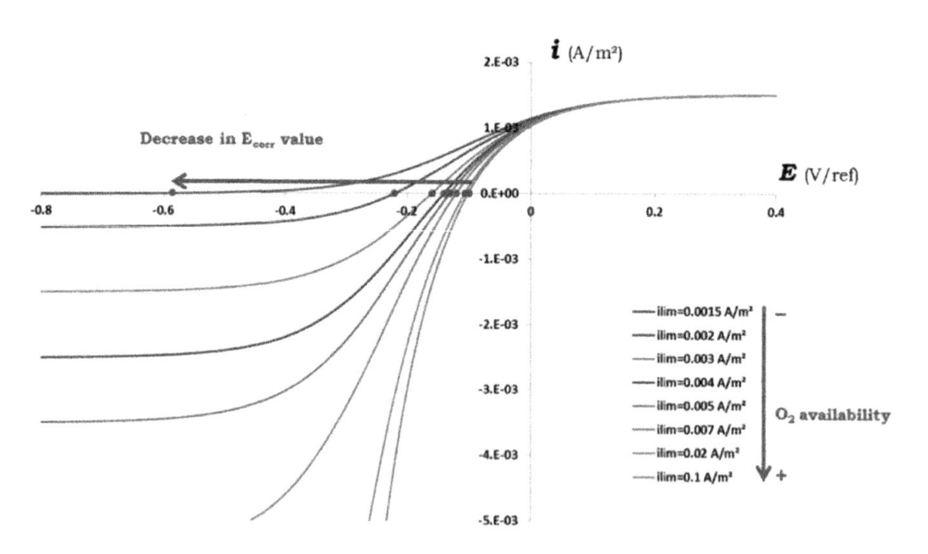

Figure 1.25. *Effect of dioxygen availability on the polarization curve of a passive steel. For a color version of the figure, see www.iste.co.uk/francois/corrosion.zip*

The dioxygen availability therefore constitutes another factor affecting the mobilized passive surface for the cathodic reaction and consequently the corrosion kinetics. Where there is an abundance of dioxygen present around the initiated anodic site, the mobilized passive surface is immediately adjacent, which implies minor ohmic drops between active and passive sites and a high corrosion rate. In the event that a corrosion site is initiated in an area difficult for the dioxygen to access, the mobilized passive surface will be delocalized to areas where O_2 is available. It should be noted that the mobilized passive surface and corrosion kinetics may moreover develop over time should the dioxygen consumption kinetics be higher than the rate at which it is renewed.

1.3.2.4.4. Limitation of the anodic process

The dissolution of iron at the anode that contributes to the formation of a layer of corrosion products may be reduced if this layer is densified. Few works exist on this subject but this assumption of very significant reduction in corrosion kinetics at the anode is put forward by Yu *et al.* [YU 15a] to explain why corrosion initiated at service crack tips (see Chapter 2) does not lead to corrosion propagation. The initial corrosion products formed lead to the formation of a rust "plug" permitted by the "expansion tank" formed by the presence of the service crack.

1.3.2.5. *Mass loss – Faraday's law*

The loss in mass of a metal subject to the corrosion phenomenon is derived from Faraday's law (equation [1.38]):

$$m = \frac{A.I_{corr}.t}{z.F} \quad [1.38]$$

with:

– m: metal mass loss [g];

– A: molar mass of the metal [g/mol];

– I_{corr} : corrosion current [A];

– t: corrosion duration [s];

– F: Faraday constant [C];

– z: metal valency.

Where the corrosion current intensity is known, this formulation of Faraday's law allows the overall mass loss to be calculated. However, as mentioned above, steel corrosion in concrete is not uniform. It is therefore preferable to reason using the corrosion current densities, i_{corr} in A/m², in order to assess the local mass losses (in g/m²). On the basis of the steel density, the mass losses can then be easily converted into thickness loss. However, as the following section demonstrates, the total corrosion current on an anodic site contributes above all to extending the surface of the attacked area. In this case, it is not appropriate to convert the total corrosion current into a loss of cross-section (extension at depth).

1.3.2.6. *Growth of active sites*

Observation of the localized-corrosion pattern of the concrete reinforcements reveals that the growth of an anodic site (corrosion spot) is always more pronounced at the surface than at depth (Figure 1.26). Thus, *the concept of pitting, as strictly defined in electrochemistry (attack at depth) does not seem to correspond to the observations made* under laboratory conditions.

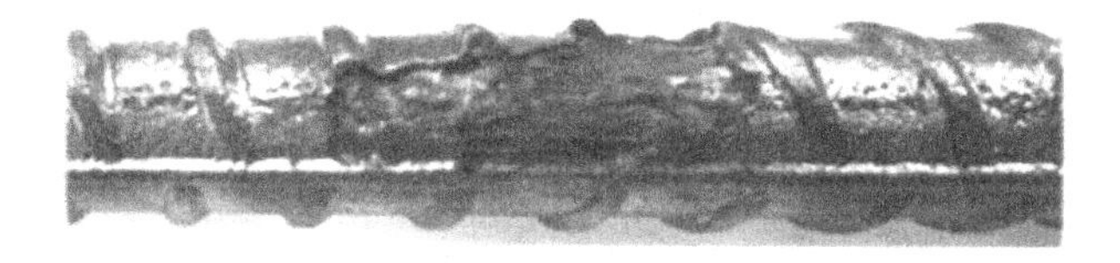

Figure 1.26. *Typical pattern of reinforcement corrosion by chlorides. For a color version of the figure, see www.iste.co.uk/francois/corrosion.zip*

A phenomenological and numerical elementary analysis enables a simple explanation of this mode of anodic site growth. Figure 1.27 illustrates the phenomenon of local depassivation resulting in the creation of a small-sized anodic site. Depassivation corresponds to the local destruction of the passive film induced by the presence of a large amount of chlorides. As soon as the active site is created, a galvanic corrosion system is formed, the anodic (iron dissolution) and cathodic (reduction of dissolved dioxygen in the solution) reactions taking place at the active and passive sites, respectively. In the following diagrams, the arrows reflect the production sites and the directions of travel of the electric charges under the effect of the electric field induced by the creation of the galvanic system:

– red arrows for positive charges (Fe^{2+} ions in concrete);

– black arrows for negative charges (OH^- ions in concrete and e^- electrons in metal).

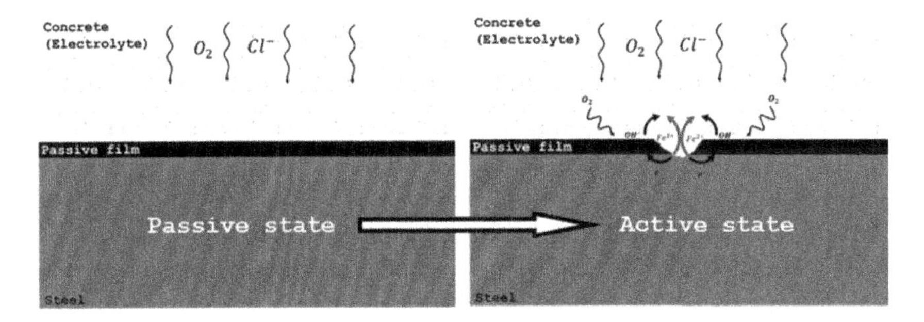

Figure 1.27. *Initiation of an active corrosion site. For a color version of the figure, see www.iste.co.uk/francois/corrosion.zip*

In the schematic representations that follow, the thickness of the arrows reflects in qualitative terms the spatial variations of the current densities produced locally.

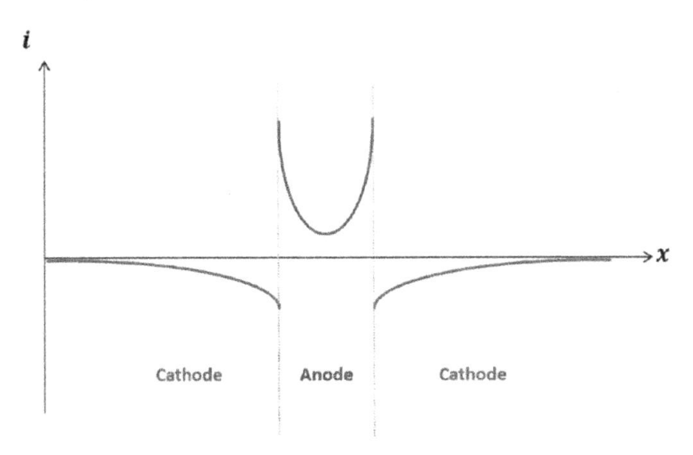

Figure 1.28. *Distribution of current densities produced on an active site and on the surrounding passive sites. For a color version of the figure, see www.iste.co.uk/francois/corrosion.zip*

A rapid calculation of the system thus established indicates that the galvanic corrosion current produced by the active site (anode) is not uniform. Figure 1.28 provides a schematic illustration of the distribution of current produced at the surface of the bar along a line crossing the anodic site. The passive sites produce a negative current (conventionally) and the active site generates a positive current. It can be noted that these currents take maximum values at the boundaries between active and passive sites. Thus, from an anodic point of view, the iron dissolution kinetics is very high at the anode/cathode boundaries and more limited at the center of the anodic site.

Figure 1.29 illustrates in qualitative terms the evolution across the surface and at depth of an anodic site. During the course of the anodic site's growth, the corrosion current density remains higher close to the boundaries between active and passive sites owing to maximum galvanic coupling. Consequently, the extension is greater across the surface than at depth. *The concept of pitting as conventionally defined in electrochemistry does not in reality concern structural steel reinforcements subjected to chloride-induced corrosion.*

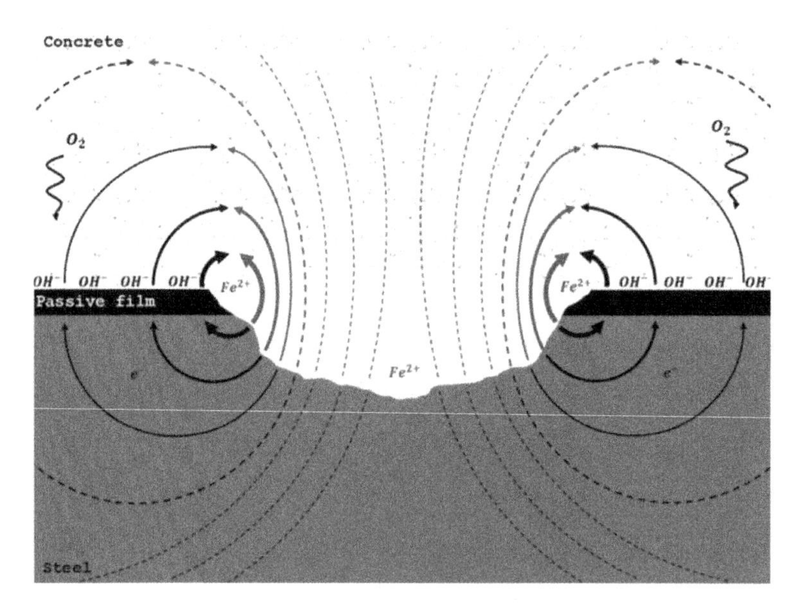

Figure 1.29. *Schematic illustration of the distribution of anodic and cathodic current densities. For a color version of the figure, see www.iste.co.uk/francois/corrosion.zip*

Thus the standard association between corrosion current and pure cross-section loss needs to be at least partially called into question. Indeed, the growth of the anodic surface is faster than its extension at depth. As the anodic surface increases, the A/C ratio increases and the corrosion current density decreases as a result. The assessment of a loss in cross-section during a structure's lifetime therefore needs to include not only the development of the anodic surface, but also the fact that the corrosion process is modified by the creation of cracks in the cover concrete, induced by corrosion (section 1.3.3). Thus, the kinetics of local loss of steel cross-section is high to begin with after initiation (small-sized spot) and slows down continuously when the anodic surface increases. From the point of view of durability calculations, this is a conservative result.

1.3.3. *Evolution of the corrosion process with corrosion duration*

The corrosion process leads to the occurrence of the first induced cracks, after a certain time period, t_c. This time period will be referred to in Chapter 4, which is dedicated to *in situ* diagnosis of corrosion, and Chapter 7, dedicated to the prediction of the service life of corroded structures.

NOTE.– The occurrence of corrosion-induced cracks, which are parallel to the reinforcement bars, leads to a progressive change in the corrosion process. Indeed, at the corrosion-induced cracks, the reinforcements are directly exposed to the environment, which leads to the development of an atmospheric corrosion mechanism, which overlaps the localized corrosion mechanism.

1.3.4. *Natural corrosion and accelerated corrosion under electrical field*

Studies and results are often found in the literature based on tests in which the reinforcements are anodically polarized, presented using the term *accelerated corrosion* under electrical field.

NOTE.– Taking into account the description of the corrosion mechanism provided in this chapter, where corrosion appears localized with the presence of anodic and cathodic areas on the reinforcements, it is clear that to impose an electric field in order to render the reinforcements anodic would not represent the mechanism of natural corrosion, or its consequences.

In this book, we will therefore only refer to results of natural corrosion that are based on long-term studies, a summary presentation of which is made in the Appendix.

2

Scale and Structural Effects on the Corrosion of Reinforced-Concrete Reinforcements

2.1. Introduction

Beyond the local corrosion mechanism described in Chapter 1, the corrosion process is influenced by the structure's implementation conditions, its dimensions, the position of the reinforcements in relation to the casting direction, the existence of several reinforcement layers and the operating conditions, particularly the mechanical service loading accompanied by the creation of cracks (known as "service cracks"). These various points are summarized using the term "scale and structural effects", and it is the influence of these effects on the corrosion of reinforced-concrete reinforcements that will be specified here.

Thus in this chapter we will discuss the influence of service cracking of reinforced concrete on the initiation and propagation of reinforcement corrosion, as well as providing a brief summary of the current recommendations concerning the controlling of crack opening in concrete structures to avoid the risk of corrosion. We will go on to discuss interface defects between the steel and concrete, linked to the significant sizing of structures (scale effects) and to mechanical loading, concrete casting direction and height. Lastly, we will explore the influence that the presence of several reinforcement layers has on the corrosion process.

2.2. Influence of cracks on the corrosion process

The presence of cracks in concrete structures is assumed to induce at least three significant effects [BEN 97]: (1) easier access to chlorides and CO_2 to depassivate the steel at the crack tip, (2) easier access to dioxygen contributing to the acceleration of the corrosion process (cathodic current), (3) they lead to physical and chemical heterogeneity at the steel-concrete interface likely to favor the corrosion process. This perception of the role of cracks, which leads one to assume that they have a decisive role in the corrosion process, does not stand up to more in-depth analysis, which will now be developed. To do so, we will provide a scientific assessment here regarding the influence of cracks on reinforcement corrosion, not presented in the exhaustive form of all works having been conducted on the subject, but rather showing that the type of experiments used, and in particular the fact that the results obtained in the short term are extended to the long term, leads to contradictory results in the literature. The effect of cracks on the ingress of aggressive agents will be studied, as will the consequences on the reduction of the initiation period and on the propagation phase. We will also discuss the variability in crack width measurement, how it evolves depending on the concrete cover, and what effect the concrete cover has on the development of corrosion in the presence of cracks.

2.2.1. *Service cracks and their opening*

Service cracking is characterized by its *visible* nature and, consequently, by its *opening*. But when crack opening is referred to, does this mean the visible opening at the surface of the concrete or the opening at the steel reinforcements? In fact, both concepts are used, depending on whether talking in terms of limitation by standards for esthetic reasons, or of durability.

The opening at the steel reinforcements is not directly accessible for measurement and can be significantly different from that measured on the surface, particularly in the case of a bending load; thus ACI 224 [ACI 01] proposes a relationship connecting crack opening on the concrete surface and the distance away from the reinforcements in order to reach its maximum value on the tensile surface.

$$wb = ws * (1/(1 + ts / h1)) \quad [2.1]$$

whereby:

wb = the most likely maximum opening at the bottom of the beam, in mm;

wb = the most likely maximum opening at the reinforcements, in mm;

ts = cover thickness as far as the center of the bar, in mm;

$h1$ = distance between the neutral axis and the reinforcements, in mm.

For the notions of limitation of crack opening for esthetic reasons, it can be assumed that this limitation concerns the maximum opening visible and measurable on the surface. However, the measurement of the opening of a crack varies significantly along the crack (see Figure 2.1) and according to the operator, the value obtained may be very different.

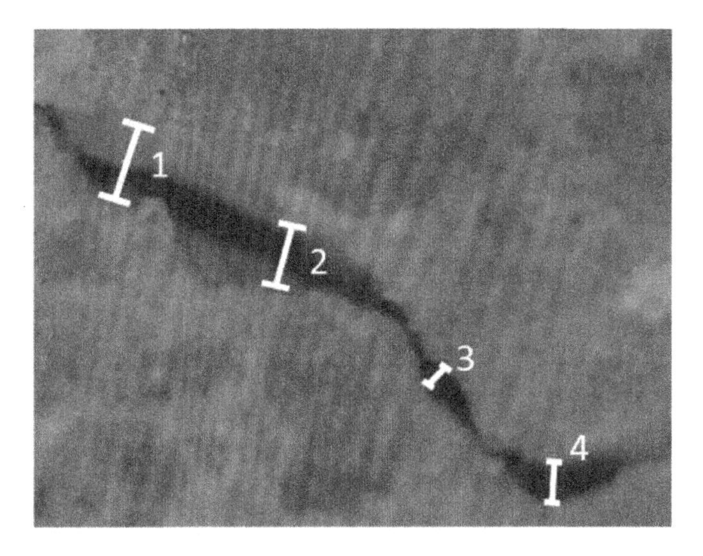

Figure 2.1. *Measurement of crack opening at concrete surface: which value should be chosen, the maximum or the average? (photo credit, Raoul François). For a color version of the figure, see www.iste.co.uk/francois/corrosion.zip*

2.2.2. Service cracks and damaging of the steel-concrete interface

Service cracking is accompanied by damage at the steel-concrete interface, which may be defined by debonding (or slipping) around the crack surfaces and by microcracking in the median area located between two consecutive cracks [GOT 71]. The formation of service cracks occurs in two phases: formation then stabilization [COM 85]. No additional crack is formed during the stabilization phase, with only the existing crack openings increasing.

The debonding area around the crack surfaces may be highlighted by a carbonation test, with 3% CO_2, the protocol of which is defined by Ghantous *et al.* [GHA 17a]. Indeed, this protocol is similar to atmospheric carbonation. Accelerated carbonation, with 50% CO_2, leads to more significant carbonation along the interface owing to carbonation shrinkage.

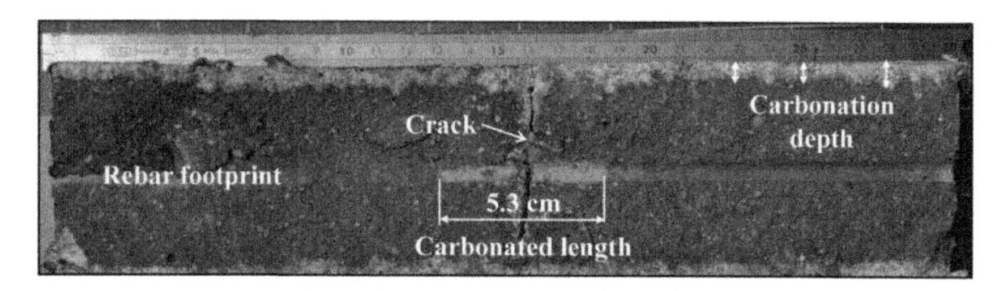

Figure 2.2. *View of the damaged area (debonding) of the steel-concrete interface accompanying the creation of a service crack, according to Ghantous* et al. *[GHA 17a]. For a color version of the figure, see www.iste.co.uk/francois/corrosion.zip*

Microcracking between two cracks occurs as a consequence of the retensioning of the concrete located between two cracks by means of the bond phenomenon. This phenomenon, known as the tension-stiffening effect, contributes to tensile stiffness for ties and bending stiffness for beams, and contributes substantially to the overall mechanical behavior of reinforced-concrete elements. However, this phenomenon leads to mechanical damaging of the concrete (internal cracks and damage to the steel-concrete interface (see Figure 2.3)) after the cracking stabilization phase [CAS 11].

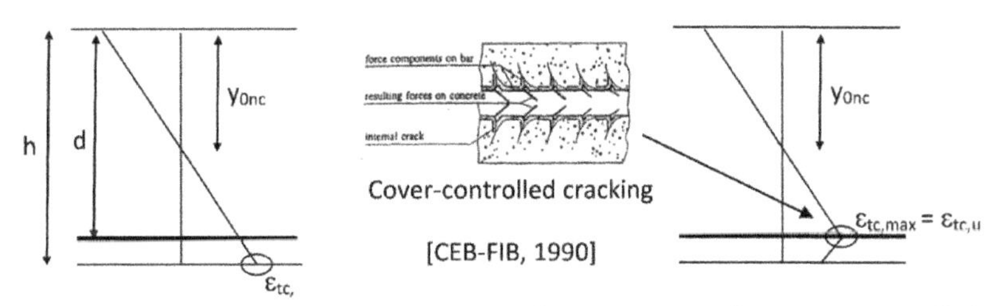

Figure 2.3. *Concrete strain prior to cracking and after cracking between two cracks, according to Castel and François [CAS 11]*

The phenomenon of internal cracking at the steel-concrete interface occurs if the strain of the tensile concrete at the reinforcements $\varepsilon_{tc,max}$ reaches the tensile concrete strain limit $\varepsilon_{tc,u}$ (Figure 2.3). From this strain limit, which depends on the cover, damage to the steel-concrete interface is obtained, possibly with the creation of microcracks breaking through to the concrete surface (see Figure 2.4).

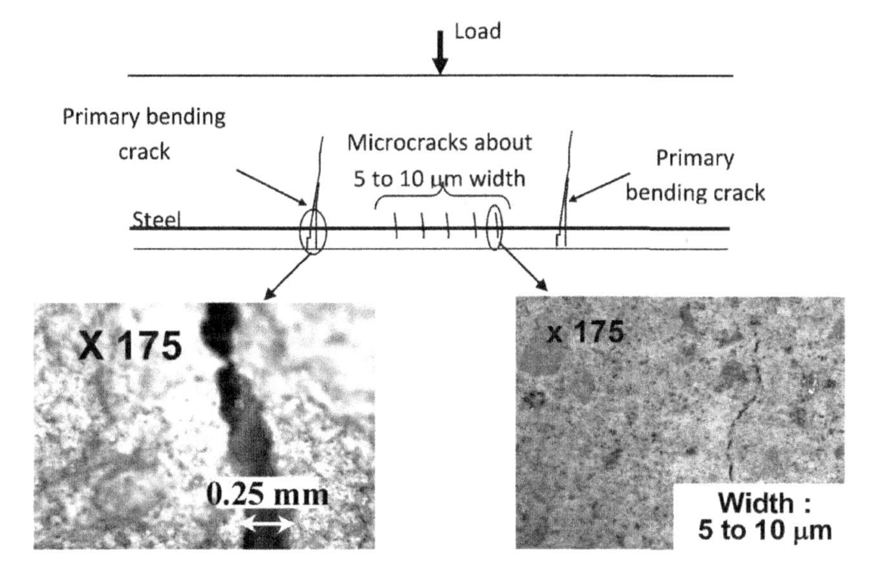

Figure 2.4. *Observation, using a video-microscope, of microcracks occurring at the concrete surface facing the reinforcements owing to the cover-cracking phenomenon, according to Castel and François [CAS 11]. For a color version of the figure, see www.iste.co.uk/francois/corrosion.zip*

2.2.3. *Service cracks and their influence on liquid, ionic and gaseous transfers*

All transportation types are favored by cracks: liquid, gaseous and ionic.

Regarding gaseous transfers, and in particular the CO_2 responsible for carbonation: Al-Ahmad *et al.* [ALA 09] and Dang *et al.* [DAN 12] demonstrate that whatever the crack opening, carbonation always occurs on the crack surfaces and that consequently, the carbonation then propagates on the damaged area of the steel-concrete interface further to the creation of a transverse crack (see Figure 2.2).

Regarding liquid transfers, water or solution containing chlorides, penetration also occurs whatever the crack opening and also leads to propagation along the damaged steel-concrete interface, as demonstrated by François and Maso [FRA 88] and Michel *et al.* [MIC 13].

The concept of a *limiting* crack opening preventing the CO_2 or the Cl^- ions from accessing the reinforcements is therefore in no way scientifically justified. Limiting of crack opening may only be associated with the renewal of aggressive substances inside the cracks. Thus, as far as ionic transfers in a saturated crack are concerned, ion diffusion is likely to occur without any restriction for an opening range beyond 40–50 microns [ISM 08], but there will likely be a restriction below this value, reducing the distribution perpendicular to the crack plane, but not the access to the reinforcement via the crack.

2.2.4. *Service cracks and their influence on the corrosion mechanism*

The phenomenon of reinforcement corrosion is always described in two phases: initiation and propagation [TUU 82]. The initiation phase corresponds to the period prior to corrosion induction and the propagation phase to the period where corrosion is active and developing. In the presence of cracks, reinforcement corrosion is always initiated at the crack tip within a few weeks, whether in the presence of chlorides or carbonation.

Indeed, it was recalled in section 2.2.3 that chlorides and CO_2 penetrate into even very fine cracks to reach the reinforcements and then propagate on the damaged area of the steel-concrete interface.

Whilst it may be clear that the start of corrosion is always initiated at the crack tip, in contrast, the continuing of the corrosion phenomenon in connection with the cracks (the propagation phase) is still a matter for debate. It is to be noted that the concept of continuing or stopping corrosion is connected to the corrosion kinetics, with the "stopping" of corrosion corresponding to very low kinetics. The fact that there is a debate in progress offers proof that the two types of result, continuing of corrosion in the cracked area and no continuing of corrosion in the cracked phase, have been found by different researchers. The conditions for obtaining these results are therefore capable of altering the conclusion in one direction or the other. The literature is therefore divided into two classes of results:

– Cracking, and crack opening in particular, accelerate the propagation of corrosion, [PET 96, SCO 07, SUZ 90, OTI 10].

– Cracking, and crack opening in particular, have no effect on the corrosion propagation phase [ARY 95, BEN 97, BEE 78, FRA 99, FRA 06a, SCH 97].

It should be noted that the majority of studies on this subject correspond to a saline environment, and these studies were conducted under accelerated conditions with an impressed current, for short periods ranging from a few weeks to a few months.

In contrast, when "natural" corrosion conditions are applied, it can be noted that the corrosion kinetics, initially very high during initiation, decreases very rapidly to a negligible value. This is the case, for example, with the results of Ghantous *et al.* [GHA 17b], which show the corrosion kinetics (initiated by carbonation) becoming negligible owing to the blockage, by the corrosion products, of the anodic area created around the crack tip at initiation (see Figure 2.5).

This reduction in corrosion kinetics at the crack tip is also observed by Yu *et al.* [YU 15a] in the case of initiation by chlorides, thus confirming the results of François *et al.* [FRA 94a, FRA 94b].

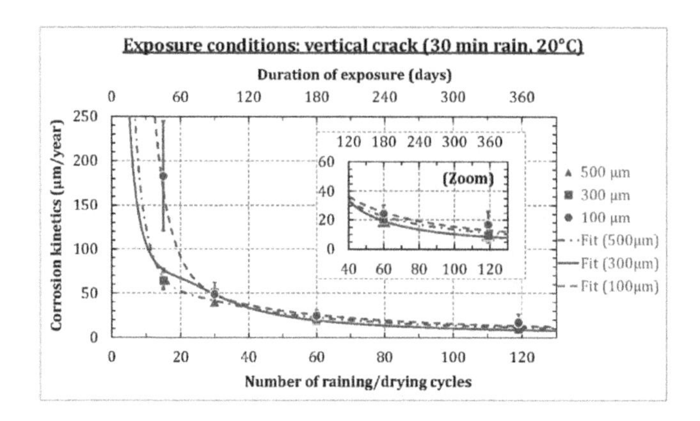

Figure 2.5. *Evolution of corrosion rate at crack tips for different crack openings, according to Ghantous* et al. *[GHA 17b]*

In conclusion, service cracks in reinforced concrete are not linked to earlier propagation of corrosion in reinforced-concrete structures. The two-phase phenomenological model introduced by Tuutti [TUU 82] therefore needs complementing by the four-phase model (incubation, initiation, induction phase and propagation) proposed by François *et al.* [FRA 94a, FRA 94b] (Figure 2.6), for cracked structures, which demonstrates that cracks do not lead to corrosion propagation, but that damage to the steel-concrete interface linked to the formation of service cracks can, however, reduce the duration of the induction phase. It should be noted that in the case of carbonation, Tuutti also demonstrates that cracks have no influence on corrosion propagation (Figure 2.7) [TUU 82].

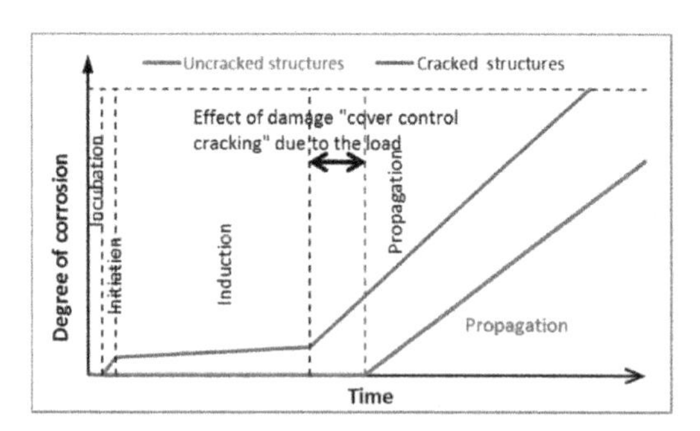

Figure 2.6. *Phenomenological models of the corrosion process in the presence of chlorides in concrete [TUU 82] and in cracked structures (François* et al. *1994) according to Yu* et al. *[YU 15a]. For a color version of the figure, see www.iste.co.uk/francois/corrosion.zip*

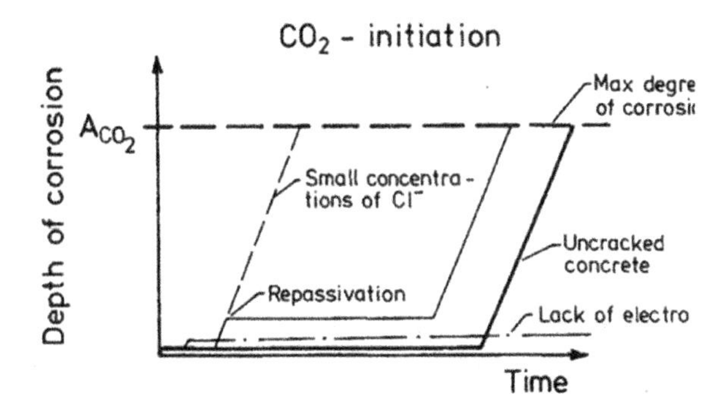

Figure 2.7. *Phenomenological models of the corrosion process in carbonated, cracked concrete according to Tuutti [TUU 82]*

NOTE.– In summary, it can thereby considered, from a corrosion perspective, that a cracked structure may be regarded as if it is not cracked, with the sole restriction that the presence of damage to the tensile concrete around the reinforcements may affect chloride penetration or carbonation [FRA 94a, FRA 94b]

2.3. Standard-based approach concerning the influence of service cracking on reinforcement corrosion in reinforced concrete

All standards include cracking control requirements. Nevertheless, the two main European and American construction standards (EuroCode 2 [EUR 92] and ACI 318 [ACI 05]) each view the effect of cracking on service life in the presence of corrosion risk differently.

Eurocode 2 considers cracking to be detrimental to durability and therefore limits its maximum opening, calculated in relation to the exposure class, to values ranging from 0.2 to 0.3 mm. The limitation is assured by controlling the spacing, diameter and quantity of the reinforcement.

Limitation (or crack control) is assured by three constraints:

– A calculated crack opening limit, w_k.

– A minimum reinforcement cross-section: this constraint is also used to assure minimum ductility on reinforced structures.

– The absence of reinforcement yielding (no inelastic deformations) at the cracks.

The ACI construction standards [ACI 05], meanwhile, consider, firstly, that the crack opening calculation leads to very significant dispersion and that by limiting this opening, at least 40% of the cracks not fulfilling this criterion are obtained [LEO 77] and secondly, that the role of cracks on corrosion is a matter of controversy.

The ACI quotes Darwin *et al.* [DAR 85] and Oesterle [OES 97], demonstrating a lack of correlation between crack openings measured at the concrete surface (within the standard range corresponding to stresses in steel for service loads) and corrosion. ACI Committee 224 [ACI 01] also notes that according to Beeby [BEE 83], there is no link between long-term corrosion and the bending-crack openings. As a consequence, no requirement is imposed as to crack opening for corrosion risk. Nevertheless, a control of the crack openings is conducted at least for aesthetic reasons, by means of a maximum spacing between the reinforcement bars.

Most international standards are in line with the Eurocode 2 approach, considering cracking to be detrimental to durability. Thus, a certain opening threshold situated within the range 0.2–0.4 mm is proposed, below which reinforcement corrosion is not influenced by the presence of cracks. There is no direct verification of these cracking thresholds, however, and it is more a matter of adapting the spacing between the reinforcement bars. (*Indeed, Eurocode 2 is the only standard to request proof of the crack opening via a calculation*).

The Australian standard, AS3600, considers cracking to be detrimental to durability and controls cracking by limiting tensile stress in reinforcements, as did the French BAEL (*Béton Armé aux Etats Limites*: limit state design of reinforced concrete) standard, [BAE 82]) used prior to Eurocode 2. This limitation depends on the diameter and spacing of the bars. The comparisons of the standards, conducted by Jenkins [JEN 09], demonstrate that the design is similar to that of the EC2 for concrete covers of less than 50 mm, but leads to crack openings significantly greater than 0.35 mm for larger covers. It should be noted that in the Australian standard, the aim of limiting steel stress is to limit the crack opening according to the environment: to 0.2 mm in saline environment, 0.3 mm for visible surfaces and 0.5 mm for non-visible surfaces [GIL 99].

Limitation of crack opening to avoid the risk of corrosion is an "obvious" parameter when the problem is not examined in detail. Indeed, as the cracks are visible, their consequences should depend on their opening. This is summed up well by Steinar Helland [HEL 12]: "*Intuitively we assume that cracked structures will deteriorate faster than un-cracked structures. However, neither the fib nor the ISO committee was able to come up with any general model to take this effect into account*".

In summary let us quote Mark Alexander [ALE 12]: "*the concept of a universal threshold crack width below which corrosion is assumed to be negligible is questionable*". In other words, limiting crack opening within the range 0.2–0.4 mm according to the aggressiveness of the environment does not allow the absence of corrosion to be guaranteed in the short or long term.

NOTE.– The approach with respect to standards therefore needs to be entirely reviewed.

2.4. Defects and damaging of the steel-concrete interface

Not only is the steel-concrete interface of great importance to the mechanical behavior of the reinforced-concrete composite via the bonding phenomenon, but evidence of its influence on durability is also progressively emerging [ANG 17b]. In particular, in the case of corrosion initiation in the presence of chlorides, it would appear that a large number of contradictory results could be explained by a failure to correctly take into account or interpret the influence of the local characteristics of the steel-concrete interface [ANG 09, ANG 17a].

Figure 2.8 summarizes the main defects that can be found at the steel-concrete interface.

The main defects of the steel-concrete interface correspond to the presence of air bubbles or entrained air, voids beneath the reinforcements due to the phenomenon of settlement and bleeding of fresh concrete, and mechanical debonding of the steel-concrete interface.

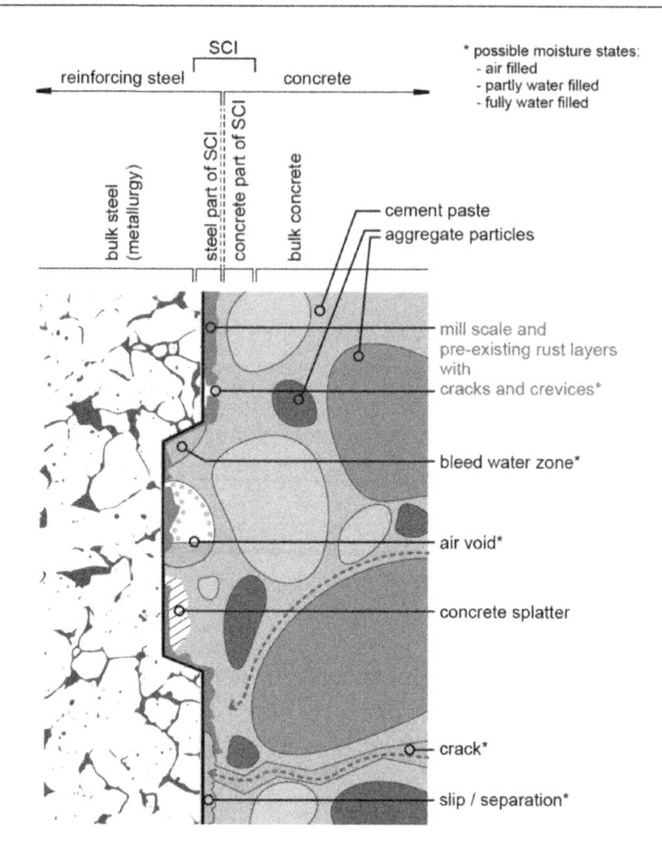

Figure 2.8. *Schematic illustration of various "defects" that may or not be present locally at the steel-concrete interface. The red dotted lines indicate preferential routes of entry for chlorides; the blue dots represent adsorbed water (for large pores only), according to Angst* et al. *[ANG 17a]. For a color version of the figure, see www.iste.co.uk/francois/corrosion.zip*

The settlement of fresh concrete and bleeding prior to the end of hardening result in lower quality concrete beneath the reinforcement bars, which remain stationary while the concrete is setting, with respect to the casting direction. This phenomenon exists for all bars, even those positioned above a shallow height of concrete, but the defect is microscopic in this case. Thus, Horne *et al.* [HOR 07] demonstrated that there was an area of greater porosity beneath the horizontal bars. This result is confirmed by Kenny and Katz [KEN 15], who compare the steel-concrete interface in the case of horizontal and vertical bars.

In contrast, with an increase in the concrete height, typically as of an order of magnitude of 150 mm [SÖY 03], macroscopic defects are obtained in the form of voids beneath the reinforcement (see Figure 2.9). In the presence of several reinforcement layers, the upper bars present more marked defects, referred to as the "top-bar effect".

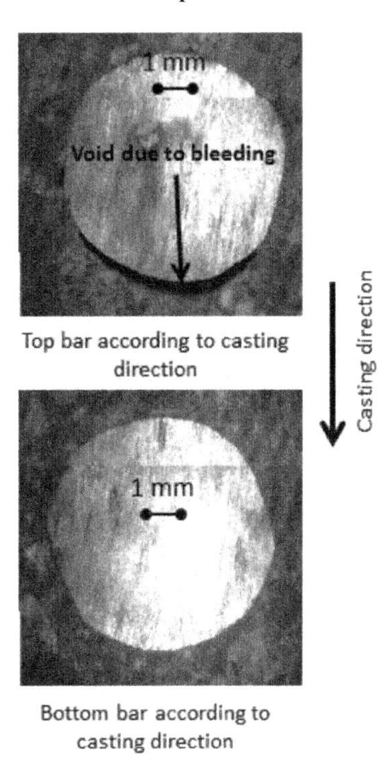

Figure 2.9. *Illustration of steel-concrete interface defects due to the bleeding phenomenon, more marked on the upper bars with respect to the casting direction [SÖY 03]*

In the presence of dense steel reinforcement, but also at the reinforcement overlapping areas and at the intersection of the horizontal and vertical bars, the appearance of voids can be observed owing to segregation or compaction problems [CEB 89].

Reinforced-concrete structures are always composed of steel reinforcements with several reinforcement layers, of varying complexity (see Figure 2.10). Steel-concrete interface defects linked to concrete casting therefore always exist.

Figure 2.10. *Steel reinforcement undergoing installation in a formwork for the reinforced-concrete shell of an industrial building (photo credit, Raoul François). For a color version of the figure, see www.iste.co.uk/francois/corrosion.zip*

2.5. Influence of "structural" (concrete casting) defects on the corrosion process: experimental results

Highlighting the scale and structural effects on the corrosion process requires structure-scale experimentations, with sustained mechanical loading and exposure conditions close to natural conditions and, of course, in the absence of so-called acceleration processes such as impressed-current systems. François *et al.* [FRA 94a, FRA 94b, FRA 94c] have been developing such experimentations since 1984 (see the Appendix) with the monitoring and application of new experimentations over time in 2008 [YU 15a, YU 15b, YU 15c] and 2011 [YU 16].

The conclusions of this long-term experimental corrosion study are as follows:

– That there is no long-term difference in corrosion between elements designed at SLS and those designed at ULS, that is, for which the design led to a limitation of the crack opening.

– That there is no correlation between long-term corrosion distribution and the initial cracks of mechanical origin (see Figure 2.11).

This result confirms those of other tests conducted under natural conditions without an impressed current [BEE 83].

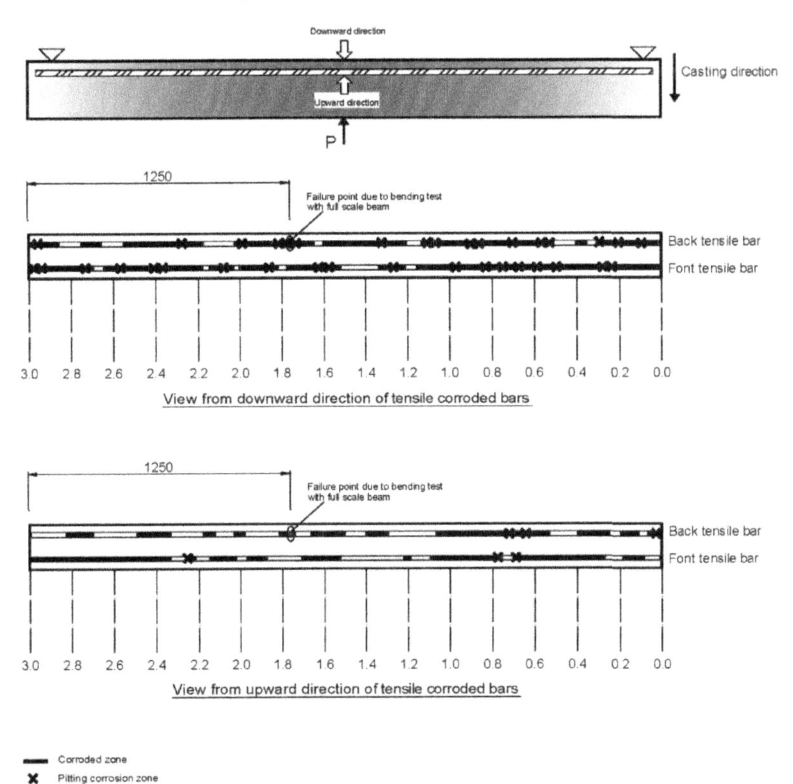

Figure 2.11. *View of corrosion progression along tensile reinforcements of a reinforced-concrete beam after 26 years of corrosion, according to Dang and François [DAN 13]. Long-term corrosion appears to have no correlation with the areas cracked by bending stress at mid-span*

Complementary experimentation was conducted within the framework of the preceding long-term research program, in order to highlight the influence of interface defects on the corrosion process. Four identical beams, C1, C2, C3 and C4, were produced. Reinforced-concrete beams C1 and C3 present the same exposure conditions, with cracked tensile lower surface, but were produced with opposite casting directions applied: tensile bars at the bottom of the formwork for C1 and at the top of the formwork for C3, respectively, with the latter thus exposed to the top bar effect. Figure 2.12(a) shows corrosion development, indicated by induced cracks, prevalent along the reinforcements presenting interface defects linked to bleeding: compressive for C1 and tensile for C3.

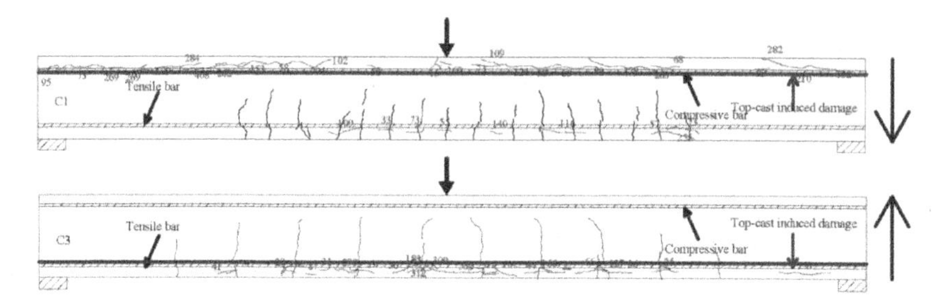

(a) Beams C1 and C3, same exposure conditions (29 months of salt spray)

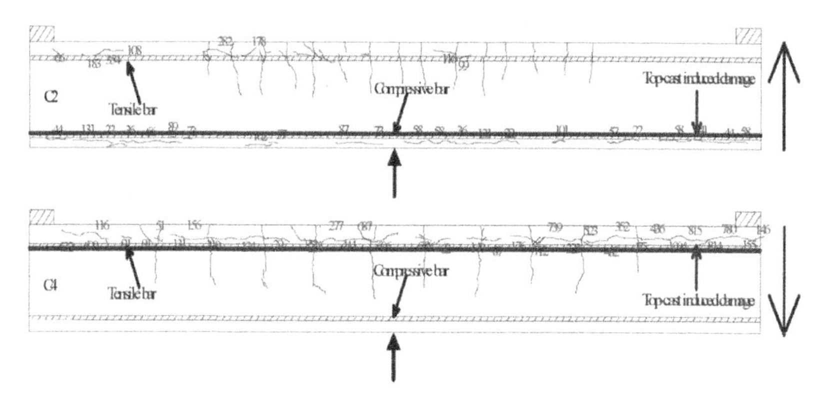

(b) Beams C2 and C4, same exposure conditions (29 months of salt spray)

Figure 2.12. *View of corrosion progression characterized by the presence of corrosion-induced cracks on four reinforced-concrete beams, which differ only in their casting direction (right-hand arrow) and exposure conditions, according to Yu* et al. *[YU 16]. For a color version of the figure, see www.iste.co.uk/francois/corrosion.zip*

Reinforced-concrete beams C2 and C4 present the same exposure conditions, with cracked tensile upper surface, but were produced with opposite casting directions applied: tensile bars at the bottom of the formwork for C2 and at the top of the formwork for C3, respectively, with the latter thus exposed to the top bar effect. Figure 2.12(b) shows corrosion development, indicated by induced cracks, along the reinforcements presenting interface defects linked to bleeding: compressive for C2 and tensile for C4.

2.6. Steel reinforcement of structures with several reinforcement layers and macro-cell corrosion process (localized corrosion)

The reinforcements of reinforced-concrete structures always present several layers connected to one another by transverse stirrups that play a mechanical role for shear resistance or simply necessary for steel-reinforcement placement reasons. An electrical connection is thus obtained between the different reinforcement layers: this electrical connection leads to a strengthening of the non-uniform nature of the reinforcement corrosion by allowing the presence of passive steel (greater concrete cover, or in less exposed areas), playing the role of cathode in the corrosion process.

Thus, on a shear stirrup connecting longitudinal reinforcements, even for very lengthy corrosion durations, non-corroded areas (the cathodic areas) and corroded areas (the anodic areas) are found, where localized attack reaches almost a 100% loss of cross-section (see Figure 2.13).

Figure 2.13. *View of the localized corrosion of a shear stirrup extracted from a reinforced-concrete beam kept under load for 28 years in saline environment: only a small part of the stirrup is corroded (anodic area), with the rest of the stirrup cathodic despite an identical chloride concentration all along the stirrup (photo credit, Raoul François). For a color version of the figure, see www.iste.co.uk/francois/corrosion.zip*

Even on the same reinforcement layer and where the chloride concentration conditions are identical all along the reinforcement bars, localized corrosion areas (corrosion pits) (see Figure 2.14) separated by cathodic areas (no corrosion) are found.

Figure 2.14. *View of localized corrosion along the tensile reinforcements of beam B1CL1 kept under load for 14 years in saline environment: Small anodic areas are separated by cathodic (non-corroded) areas despite an identical chloride concentration all along the reinforcement bars, according to Zhang* et al. *[ZHA 09b]. For a color version of the figure, see www.iste.co.uk/francois/corrosion.zip*

The initially localized corrosion leads to the creation of corrosion cracks due to the tensile stresses in the concrete, generated by the corrosion products presenting a greater volume than the Iron. The creation of these corrosion cracks leads to the progressive generalization of the corrosion through a change in environment around the steel reinforcements. Indeed, the corrosion cracks coincide with the reinforcements and authorize access to aggressive agents, to dioxygen, all along the crack. The development of corrosion cracks leads in time to the delamination of the concrete cover, which then enables *atmospheric* corrosion (Figure 2.15), which adds to the localized corrosion that can still exist in other parts of the structure.

Figure 2.15. *View of the corrosion of beam B2CL2 (corrosion duration: 26 years), which becomes generalized along the tensile reinforcements owing to the presence of corrosion cracks: photo credit, Raoul François. For a color version of the figure, see www.iste.co.uk/francois/corrosion.zip*

2.7. Conclusion

Reinforcement corrosion in reinforced concrete cannot be comprehended without taking into account the structural and scale effects present in reinforced-concrete structures.

NOTE.– The most significant factor influencing the corrosion process relates to placement defects appearing at the steel-concrete interface owing to the bleeding phenomenon. This defect is accentuated by the height of the bars in the formwork and is referred to as the "top-bar effect".

Another significant factor is the steel-reinforcement density of reinforced-concrete elements in electrical continuity, which leads to an amplification of the intrinsically non-uniform nature of corrosion in reinforced concrete with the presence of *cathodic areas* in the core of the reinforced elements. This aspect explains, in particular, why corrosion due to carbonation is also a non-uniform corrosion phenomenon, even if the anodic areas are larger than that due to chloride-induced corrosion.

NOTE.– On the contrary, service cracking visible on all reinforced-concrete structures is not, in itself, a parameter that influences the corrosion process. This is not the case, however, in the event that it is coupled with steel-concrete interface defects.

Beyond the presence of visible service cracks, mechanical loading leads to damaging of the steel-concrete interface, as well as of the tensile concrete between cracks. These phenomena reduce the time required for corrosion to propagate.

Figure 2.16 summarizes these conclusions. The first-order parameter corresponds to the steel-concrete interface defects. If these are coupled with mechanical loading, this gives a very significant reduction in the time period prior to propagation. The presence of mechanical loading alone simply leads to a minor reduction in the induction phase prior to propagation.

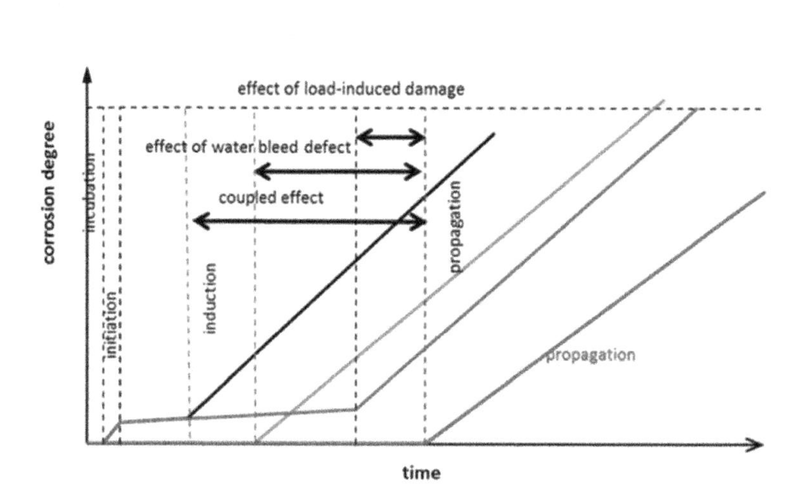

Figure 2.16. *Phenomenological models of the corrosion process in the presence of structural defects: interface defects due to bleeding, mechanical defects due to loading and coupling between the two defect types. For a color version of the figure, see www.iste.co.uk/francois/corrosion.zip*

3

Which Parameter to Quantify Corrosion Intensity?

3.1. Introduction

The corrosion process is always described as an increase in the *degree of corrosion* as a function of time. For example, the universally-quoted Tuutti model [TUU 82] (see Figure 3.1) uses the *depth of corrosion* parameter as the ordinate.

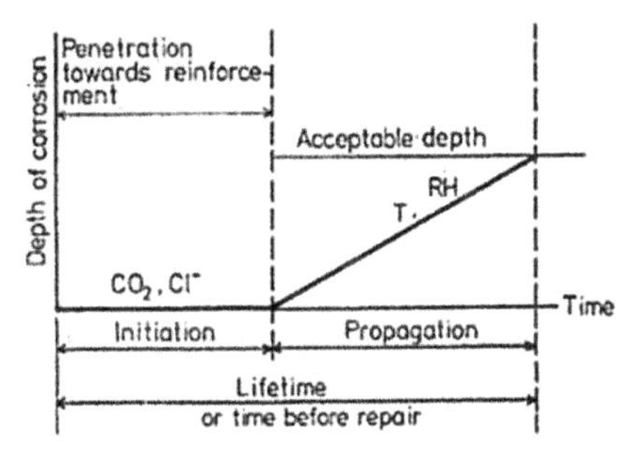

Figure 3.1. *Phenomenological model of reinforced-concrete reinforcement corrosion, according to Tuutti [TUU 82]*

Corrosion depth is local by nature, and this is even truer for reinforced-concrete corrosion, which is initially localized. Whilst Tuutti's model is a global phenomenological model, it expresses a local parameter corresponding to an anodic area of the process of non-uniform corrosion of steel reinforcements in reinforced concrete. So which is the parameter hidden behind this development in the degree of corrosion as a function of time?

Is the degree of corrosion defined with respect to the entire structure, to an element of the structure (beam, column, etc.), to part of the structural element (tensile bars, shear stirrups, etc.), to the cross-section of a structural element, or to the cross-section of each reinforcement? In other words, what is the scale of definition of the degree of corrosion?

This is an essential question in that, as mentioned on several occasions (Chapters 1 and 2), steel corrosion in reinforced concrete is mainly localized with a prevalent macro-cell (localized corrosion) rather than micro-cell (uniform corrosion) effect. As a consequence, both anodic and cathodic areas can be found on the same bar (see Figure 3.2).

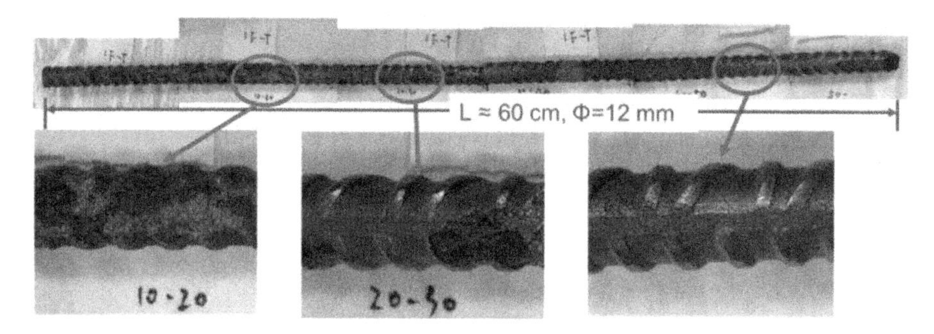

Figure 3.2. *Illustration of non-uniform corrosion along a reinforcement of a reinforced-concrete beam: beam Bs04, aged 27 months, according to Yu* et al. *[YU 15c]. For a color version of the figure, see www.iste.co.uk/francois/corrosion.zip*

Yet the presence of an anodic area on a bar does not necessarily mean that there is homogeneous corrosion across the whole perimeter (see Figure 3.3). This observation clearly confirms that *corrosion referred to as accelerated under electrical field (Chapter 1), which leads to a continuous anodic area along the steel reinforcements, cannot be representative of actual corrosion*. This observation also clearly indicates that the concept of an overall degree of corrosion makes no physical sense, as it corresponds to the average of the cross-section losses on anodic areas (where there is indeed a cross-section loss) and cathodic areas (where there is no cross-section loss).

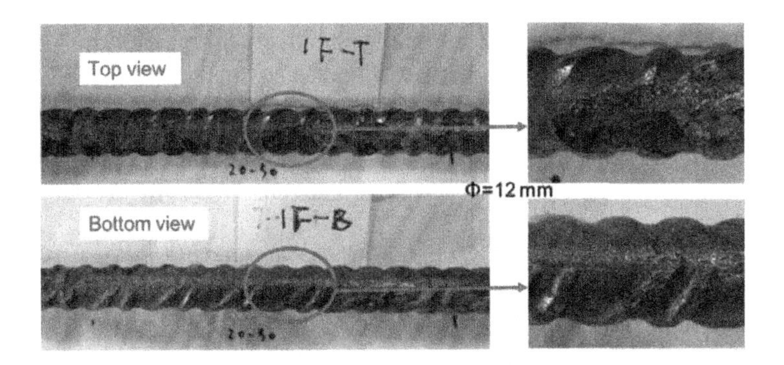

Figure 3.3. *Illustration of non-uniform corrosion around the perimeter of a reinforcement of a reinforced-concrete beam: beam Bs04, aged 27 months, according to Yu* et al. *[YU 15c]. For a color version of the figure, see www.iste.co.uk/francois/corrosion.zip*

Nevertheless, from a civil engineer point of view, while it is important to know the maximum local steel cross-section losses and their localization within the structure's geometry, as they will have an influence on the bearing capacity, it is also useful to obtain information on the generalized cross-section losses along the reinforcements, which will influence the bending stiffness and deflections in service.

In this chapter we will therefore be defining a *degree of local corrosion* (*LC* on a millimeter scale and a *degree of generalized corrosion* (*GC*)) on a centimeter scale, for a given reinforcement.

3.2. Degree of local corrosion, defined as the loss of cross-section on a reinforcement (localized corrosion, LC)

Having chosen to define a millimetric local degree of corrosion for a given reinforcement, the dimension on which this degree of corrosion is to be calculated (1 mm, 5 mm, 10 mm, etc.) then needs to be chosen, as well as how it is to be expressed: is it a pitting depth, a diameter loss or a cross-section loss, and expressed as a percentage of their initial values or as an absolute value?

Loss in diameter or pitting depth may be measured using a sliding caliper, but cross-section loss results from a measurement of mass loss, due to corrosion, on a given length of reinforcement with respect to its initial mass.

As corrosion is not homogeneous on a reinforcement, the diameter loss measurement indicates more a corrosion depth than an actual change in diameter (see Figure 3.4), which leads to the overestimation of the reinforcement cross-section loss for a structural recalculation.

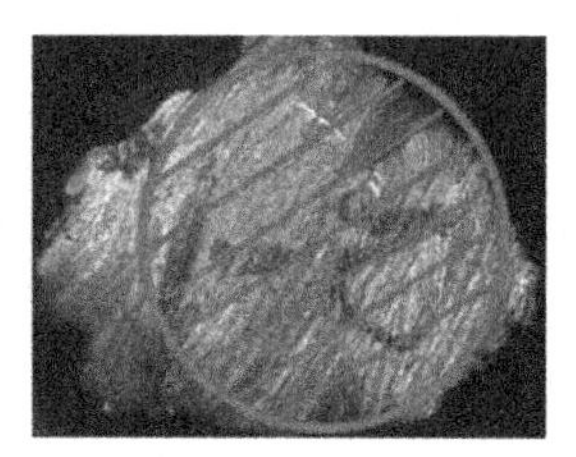

Figure 3.4. *Illustration of the residual cross-section of a corroded reinforcement derived from the measurement of the residual diameter, according to Zhu* et al. *[ZHU 17a]. For a color version of the figure, see www.iste.co.uk/francois/corrosion.zip*

Moreover, the quantification of cross-section loss based on a gravimetric measurement requires a minimum length of reinforcement, which needs to be cut out in order to be weighed: a length of less than 5 mm is hard to envisage given the inaccuracies on the parallelism of cut-out faces and on the cut-out process itself (see Figure 3.5). Reinforcement cross-section loss based on a gravimetric measurement thus appears as an average value, nevertheless for a short bar length, which makes it a local parameter.

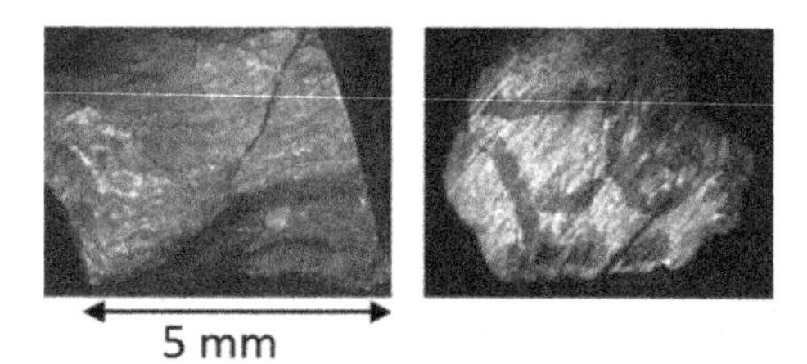

Figure 3.5. *Illustration of the residual mass measurement of a corroded reinforcement, cut into 5-mm-long coupons, according to Zhu* et al. *[ZHU 17a]. For a color version of the figure, see www.iste.co.uk/francois/corrosion.zip*

Loss of diameter or depth of corrosion may be measured continuously on a corroded reinforcement whereas cross-section loss resulting from a gravimetric measurement is inevitably discrete on pitches that can depend on the corrosion pattern along the reinforcement. Indeed, it is not necessary to cut out coupons every 5 mm in length in the absence of very local pits.

As an illustration, let us present Figures 3.6 and 3.7, respectively, the measurements of cross-section loss deduced from a gravimetric analysis and a residual diameter measurement for a piece of reinforcement extracted from a beam, B2CL2, aged 26 years [ZHU 17a]. For the gravimetric measurement, the length of the reinforcement portions varies according to the corrosion pattern and this length corresponds to the segment drawn in Figure 3.6, which presents a minimum length of 5 mm and a maximum length of 15 mm.

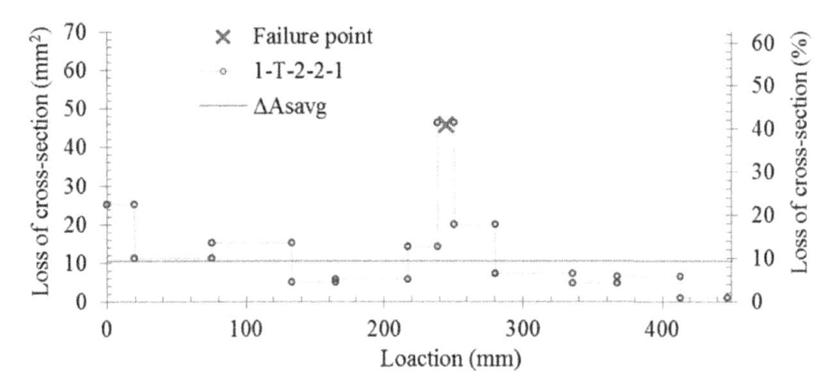

Figure 3.6. *Evolution of steel cross-section loss due to corrosion along the reinforcement portion 1-T-2-2-1 calculated by gravimetric measurement, according to Zhu* et al. *[ZHU 17a]. For a color version of the figure, see www.iste.co.uk/francois/corrosion.zip*

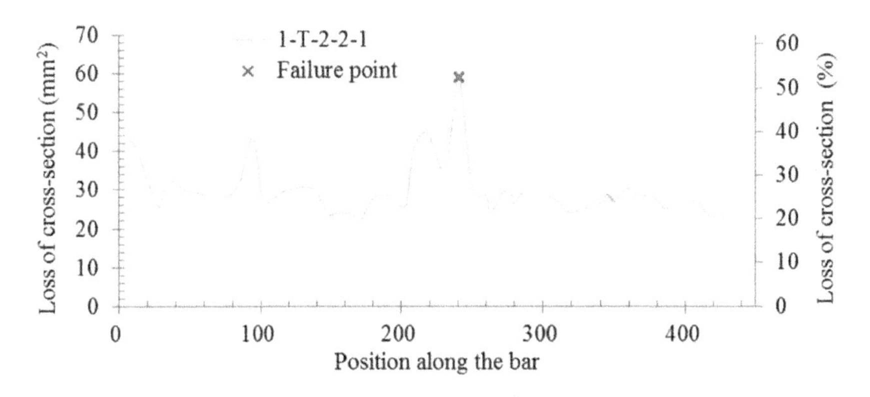

Figure 3.7. *Evolution of steel cross-section loss due to corrosion along the reinforcement portion 1-T-2-2-1 calculated by measuring the residual diameter, according to Zhu* et al. *[ZHU 17a]. For a color version of the figure, see www.iste.co.uk/francois/corrosion.zip*

For the diameter loss measurement (Figure 3.7), it is necessary to take into account the fact that reinforced-concrete reinforcements feature ribs or reliefs (deformed reinforcements) and that the nominal diameter does not exactly correspond to the diameter measured between ribs or reliefs. The measurements are therefore carried out between the ribs (where they still exist in the corroded state).

Each peak shown in Figure 3.7 represents a corrosion pit; only the highest peak has been measured by the gravimetric method (Figure 3.6). The maximum cross-section loss calculated by gravimetric measurement stands at 46.3 mm^2 (i.e. around 41% of the initial cross-section), whilst that calculated by measuring diameter losses leads to a value of 59.25 mm^2 (i.e. around 52.4% of the initial cross-section). The relative deviation between the two maximum losses is greater than 20% despite the straightforward measurement conditions applied in a laboratory situation. Thus, the diameter loss measurement leads to the cross-section loss being overestimated. Inversely, it can be assumed that the gravimetric measurement leads to an underestimation as it is averaged out over several mm; the actual value is situated somewhere between the two and is no doubt closer to the value resulting from the gravimetric method.

It is possible to conduct more accurate measurements of the cross-section losses by 3D scan. This work is shown as an example on a reinforcement 12 mm in diameter, extracted from beam C1 [YU 16], aged 28 months. Figure 3.8 shows the analyzed surface of a segment (T1) 80 mm in length and, an example of the measurement of the actual residual cross-section, which is 107.87 mm^2, corresponding to a cross-section loss of just over 5 mm^2 with respect to the nominal cross-section.

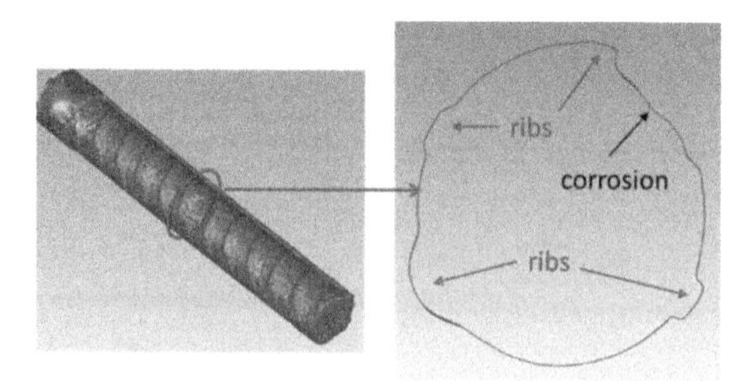

Figure 3.8. *3D scan of segment T1 of corroded reinforcement C1-Tensile bar-6_Back and of the cross-section at a local cross-section loss. For a color version of the figure, see www.iste.co.uk/francois/corrosion.zip*

Figure 3.9 shows the evolution of mass loss all along a tensile reinforcement of beam C1. The sample analyzed by 3D scan is identified in Figure 3.9; it corresponds to the central part of the bar with an average cross-section loss of 7.5 mm^2 over a length of 80mm. Figure 3.9 also presents the evolution of cross-section loss calculated all along segment T1 by 3D scan, as shown in Figure 3.8. On the right-hand part of Figure 3.9 it can be noted that the ribs (or reliefs) of the bar are very visible with their periodic signature and an amplitude of around 4 mm^2, and that the cross-section loss fluctuates around an average value of 7.5 mm^2, measured across the entirety of sample T1, varying from 0 to 17 mm^2.

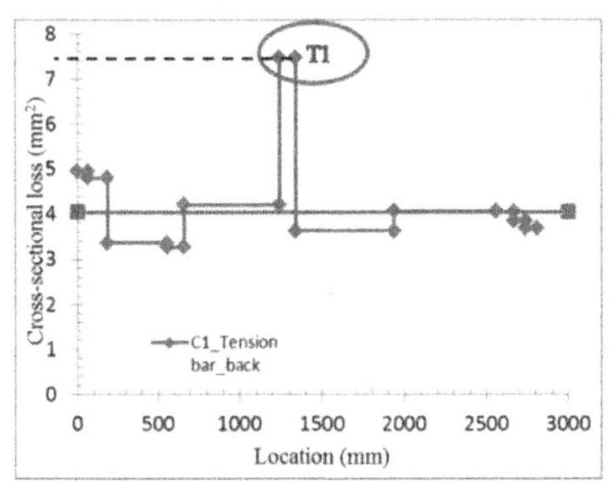

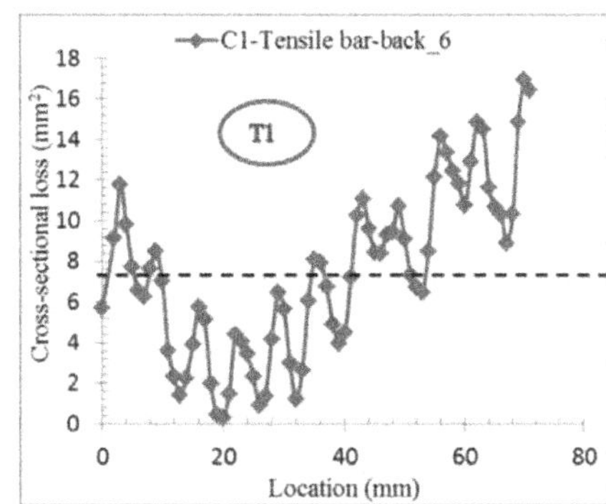

Figure 3.9. *Evolution of cross-section loss along the "tensile-back" bar of beam C1, to the left, and close-up view of part T1, 80 mm in length, on the right, analyzed via 3D scan. For a color version of the figure, see www.iste.co.uk/francois/corrosion.zip*

Figure 3.9 enables confirmation of the relevance of the gravimetric analysis with an average cross-section loss on segment T1, which corresponds to the average local cross-section losses measured by 3D scan. It can also be noted that a minimum pitch of 5 to 10 mm in the gravimetric measurements enables relevant information to be obtained on local cross-section loss, breaking away from the influence of ribs (or reliefs) of deformed reinforcing steel.

It should be noted that these measurements confirm the specific nature of the corrosion pits obtained in reinforced concrete: relatively extended in length rather than at depth.

3.3. Degree of corrosion defined as the average loss of cross-section on a reinforcement (generalized corrosion, GC)

Local losses of cross-section on reinforcements will influence the bearing capacity of the structures, provided that they are associated with significant stress areas. Nevertheless, whilst a corrosion pit changes the tensile strength of a steel bar, it does not modify its longitudinal stiffness; the latter requires generalized cross-section loss along the reinforcement. Moving on to the structural element, a beam for example, the bending stiffness, for example, is influenced by the behavior between service cracks, which depends on the reinforcement cross-section but also on the bond phenomenon. The average corrosion parameter useful in this case appears to be the average value on the spacing between two service cracks, which is of a scale of magnitude of 100 to 200 mm. The dimension chosen to calculate the average corrosion has an influence on the result: Figure 3.10 provides an illustration of this. The cross-section loss, for a reinforcement extracted from a beam, Bs02, aged 36 months, averaged out across lengths ranging from 25 mm to 220 mm, varies from 34% to 11.5%. It can be noted, however, that the difference becomes only slight as soon as a length of 120 mm is exceeded. Thus, the average corrosion value, even in the presence of significant pitting, remains within an acceptable range for the interval 100-200 mm, taking account of inaccuracies in the measurement of *corrosion* loss. This order of magnitude of 100-200 mm will thus be obtained to quantify generalized corrosion (GC).

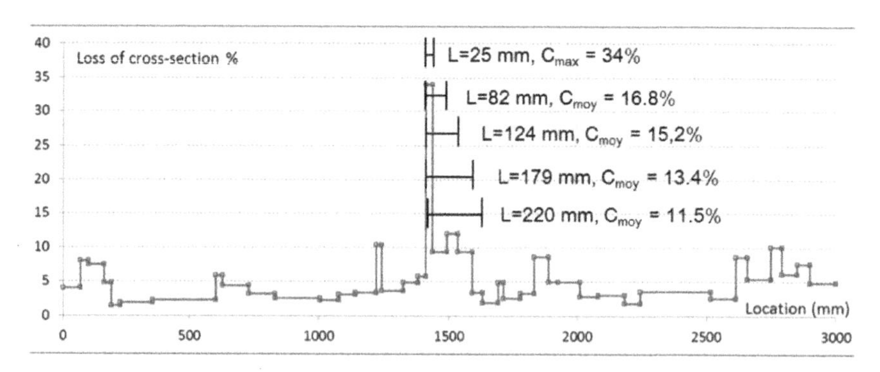

Figure 3.10. *Evolution of local cross-section (LC) loss, measured by gravimetric analysis, along a tensile bar of beam Bs02, aged 36 months, and evolution of average loss (GC) calculated based on the local loss for different bar length, according to Yu* et al. *[YU 15b]. For a color version of the figure, see www.iste.co.uk/francois/corrosion.zip*

The average loss of *corrosion*, calculated across the entire length of the reinforcement of beam Bs02 presented in Figure 3.10 is 4.5%. This information is of very little interest with respect to the mechanical behavior of the beam. It is, however, useful in order to calculate one of the expressions of the pitting factor, which is the ratio of maximum local corrosion at the most intense pit to average corrosion.

3.4. Pitting factors: pf_s, pf_g, pf_{lc}

Pitting factor can have several different definitions: it is, for example, the relationship between the maximum local corrosion of the most intense pit and the average corrosion over all or part of the structure. With this definition, the pitting factor named pf_s provides information on the non-uniform nature of the corrosion and on the capacity to predict the service life of a structure based on an average corrosion. Once again here the question is raised as to on which perimeter of the structure the average corrosion is defined. In the case presented in Figure 3.10, the pitting factor pf_s calculated on a tensile reinforcement of beam Bs02 equals 7.5. Throughout the remainder of this section we will continue to reason in terms of a structural element (typically a beam) and for a reinforcement function (or type) with respect to the mechanical function: tensile reinforcements, for example.

A second possible definition is a local geometric coefficient, pf_g, linked to the intensity of the cross-section loss due to corrosion compared to a homogeneous loss on the reinforcement cross-section. For this, an idealized pit shape needs to be defined: we suggest using a triangular pit with an angle of 2φ, which we will also refer to in Chapter 7, concerning service life prediction. In this case, the pitting factor is defined as the ratio of the cross-section loss based on the residual diameter obtained with the pitting to that corresponding to homogeneous corrosion (Figure 3.11) for the same amount of iron lost.

For this pitting geometry, a pitting factor of around 3 is obtained. *This geometric pitting factor will be used to distinguish between chloride-induced corrosion with a value of 4, and that induced by carbonation with a value of 1.*

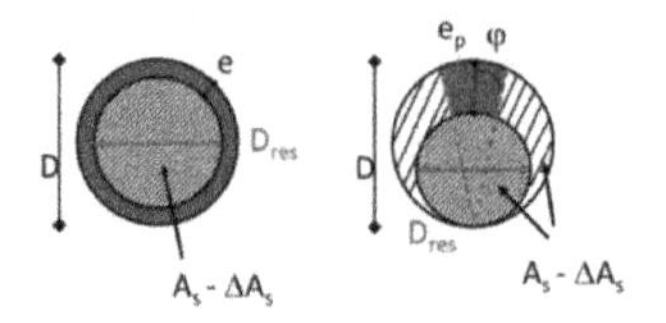

Figure 3.11. *Corrosion pitting and corresponding residual cross-section in comparison with homogeneous corrosion. For a color version of the figure, see www.iste.co.uk/francois/corrosion.zip*

A third possible definition is ratio pf_{lc}, between the local corrosion (LC) and generalized corrosion (GC) over a short distance of the order of magnitude of the spacing between the service cracks, typically within the interval 100-200 mm. This ratio will then vary according to the corrosion duration, thus at the start of the, essentially localized, corrosion process, values comparable to that of the geometric factor, pf_g, need to be found. Indeed, we verify that in the case of Figure 3.10, the pitting factor pf_{lc}, calculated at the location where corrosion is at a maximum on the tensile reinforcement of beam Bs02, is around 3.5.

In Chapter 7 we will also define another coefficient, characterizing the pitting, which is a useful form factor in predicting the ductility losses induced by pits.

3.4.1. *Can localized corrosion (LC) be described by a statistical law?*

Corrosion intensity always appears to be stochastic in nature and we might wonder whether this follows statistical models. Yu *et al.* [YU 15c] have conducted statistical analyses on beams Bs02, Bs03 and Bs04, aged several months, and Zhu and François [ZHU 16b] on a beam, B2CL3, aged 28 years.

Zhu and François [ZHU 16b] plotted (see Figure 3.12) the local corrosion frequencies (segment 5mm long) of the two tensile reinforcements of beam B2CL3 (28 years). The Shapiro-Wilk statistical test was applied to the data, and demonstrates that:

– Corrosion pattern does not follow a normal distribution.

– The presence of several areas presenting very high levels of local corrosion (>50% cross-section loss).

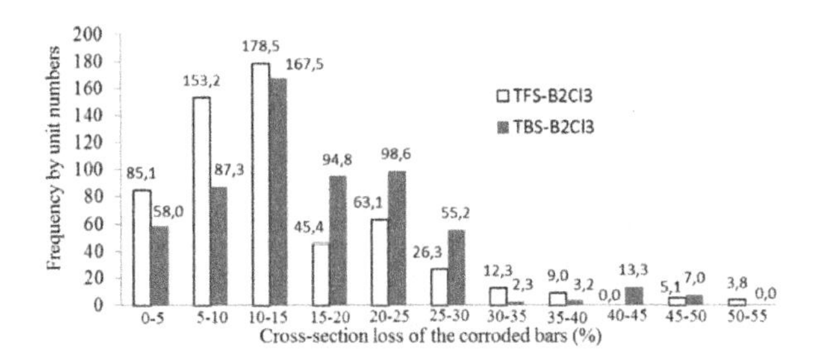

Figure 3.12. *Histogram presenting the distribution of local cross-sectional losses due to corrosion for the two tensile reinforcements of beam B2CL3, according to Zhu and François [ZHU 16b]. For a color version of the figure, see www.iste.co.uk/francois/corrosion.zip*

Yu *et al.* [YU 15c] plotted (see Figure 3.13) the local corrosion frequencies (segment of 10 mm) of the two tensile reinforcements of beams BS02 (36 months), BS03 (19 months) and Bs04 (27 months). Here too, the Shapiro-Wilk test demonstrates that the different corrosion distributions do not follow a normal distribution. Moreover, despite the three beams having the same geometry, the same concrete, the same aging conditions and despite them each having been kept under load, there is no similarity in their corrosion distribution (Figure 3.13).

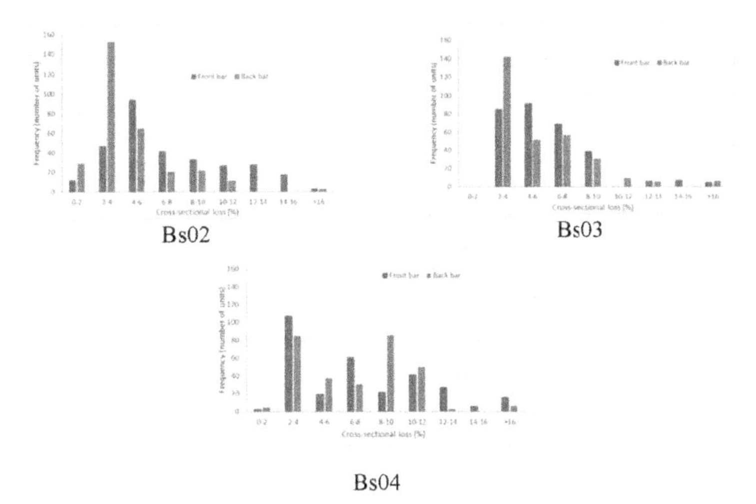

Figure 3.13. *Histogram presenting the distribution of local cross-sectional losses due to corrosion for the two tensile reinforcements of beams Bs02, Bs03 and Bs04, according to Yu* et al. *[YU 15c]. For a color version of the figure, see www.iste.co.uk/francois/corrosion.zip*

Thus the experimental results show that corrosion distribution along a structural element does not follow a normal statistical distribution. It is therefore not possible to stop at studying one part of the structure in order to deduce possible corrosion across the entire structure. Nor is it possible to reproduce the results obtained on a structure in order to generalize across other, in principle identical, structures.

3.4.2. *Field of variation of pitting factor (pf_s)*

In its conventional definition, the ratio of maximum local cross-section loss to average loss, pitting factor pf_s represents the localization of the corrosion. In principle, it should be higher at the start of corrosion, which is produced with high ratios between cathodic surface and anodic surface. Indeed, the creation of corrosion cracks enables the corrosion to develop along these cracks according to a generalized process, which thus reduces the cathodic surfaces, over time. Figure 3.14 shows the evolution in pitting factor with respect to the average loss of cross-section due to corrosion calculated along the entire length of each tensile reinforcement of beams B2CL2, B2CL3, B2CL1, B1CL1 and Bs02, Bs03, Bs04, according to Yu *et al.* [YU 15c]. A tendency for the pitting factor to reduce with respect to the average corrosion percentage can indeed be noted, as well as a decrease in dispersion between the different bars.

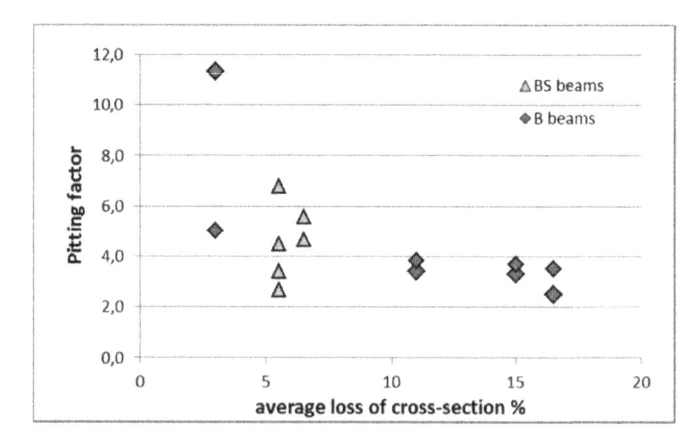

Figure 3.14. *Evolution of pitting factor with respect to the average loss of cross-section along the length of a tensile reinforcement for beams B2CL2, B2CL3, B2CL1, B1CL1 and Bs02, Bs03, Bs04, according to Yu* et al. *[YU 15c]. For a color version of the figure, see www.iste.co.uk/francois/corrosion.zip*

By plotting pitting factor pf_s for the same tensile bars of beams B2CL2, B2CL3, B2CL1, B1CL1 and Bs02, Bs03, Bs04, with respect to the corrosion duration (see Figure 3.15), a tendency can be noted for the pitting factor to decrease according to the corrosion duration, which is consistent with the corrosion gradually becoming less localized. There can, however, be a significant scatter on certain points. In all cases, including for very significant corrosion durations of almost 30 years, we always retain a pitting factor that is above 2. For shorter durations that can correspond to the date of the first appraisals with a view to the diagnosis and recalculation of the structure, a pitting factor can be observed that varies between 2 and 7. This signifies that an estimation of the structural capacity based on an average estimation of the corrosion can overestimate the residual bearing capacity within the same variation range as that of the pitting factor (2 to 7), which is not an acceptable level of accuracy. It is therefore necessary to operate a diagnosis that enables local assessment of the degree of corrosion along the structure.

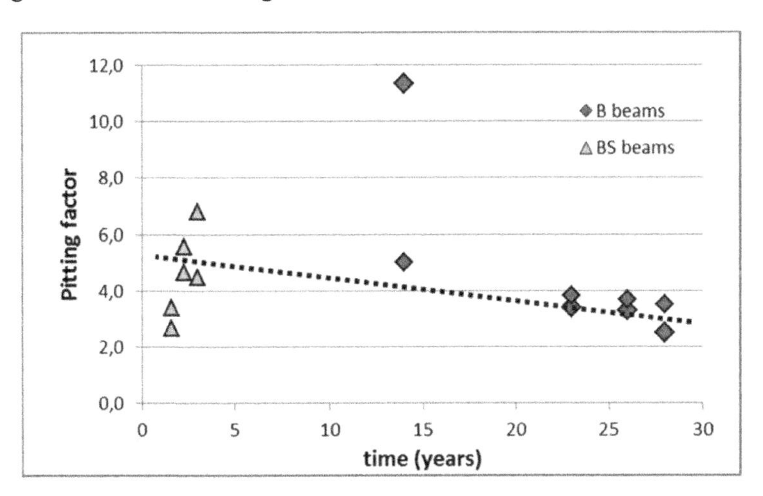

Figure 3.15. *Evolution of pitting factor with respect to corrosion duration for beams B2CL2, B2CL3, B2CL1, B1CL1 and Bs02, Bs03, Bs04, according to Yu* et al. *[YU 15c]. For a color version of the figure, see www.iste.co.uk/francois/corrosion.zip*

The other definitions of pitting factor will be useful to establish the corrosion diagnosis approach based on corrosion-induced crack opening, as presented in Chapter 4 below.

3.5. Conclusion: corrosion quantification

No correlation appears possible between the maximum corrosion measured at the pits and the average corrosion on a structural element, let alone on the entire structure. It is therefore necessary to study the corrosion in the whole structure, describing local cross-section (LC) losses, in order to be able to predict the residual mechanical behavior of a corroded structure. The gradual generalization of the corrosion process, caused by the occurrence of cracks induced by corrosion, leads to cross-section losses all along the reinforcements. It is necessary to quantify these in order to comprehend the effects of corrosion on deflections in service of corroded structures. This quantification will be performed at intervals of the order of 100-200 mm and will be referred to as generalized corrosion (GC).

The overall ratio, pf_s, between the maximum local loss of cross-section and the average loss, known as the pitting factor, is very high and greatly dispersed at the start of the corrosion process. Moreover, it remains significant (>2) for very long corrosion durations (> 30 years). It is therefore necessary to be able to quantify the local corrosion (LC) in a structure in order to predict its residual mechanical behavior.

NOTE.– The geometric ratio, pf_g, between local loss (LC) at a pit and the equivalent, "homogenized" loss on the perimeter will be used to calculate the occurrence of the first cracks induced by corrosion, presented in Chapter 4 below. A value of $pf_g = 4$ will be retained for chloride-induced corrosion and a value of $pf_g = 1$ for carbonation-induced corrosion throughout the remainder of this book.

The ratio, pf_{lc}, between local loss (LC) and generalized loss (GC), depends on the corrosion duration. Its quantification is at present impossible, other than by destructive diagnosis.

4

In situ Corrosion Diagnosis

4.1. Introduction

The aim of this chapter is to present the different possibilities for establishing a state-of-corrosion diagnosis. We will not be referring to the determining of causes of corrosion linked to carbonation depth or to a certain rate of chlorides present in the vicinity of the steel reinforcements, but to the measurement of parameters linked to corrosion development.

The corrosion diagnosis based on non-destructive testing has become a major expectation in the management of reinforced-concrete built heritage. The desired objectives are to reveal the existence of corrosion and quantify its extent and its evolution over time.

For an on-site corrosion diagnosis, it is necessary to have in-depth knowledge of the position and diameter of the reinforcements, as well as the localization of the anodic and cathodic areas. The two major electrochemical techniques most commonly used today (reinforcement electrode potential measurement and polarization resistance measurement) are presented in this chapter but do not allow a quantitative corrosion diagnosis to be established, in particular in the case of measurements giving access to the corrosion current, which are based on the generally inexact assumption of uniform corrosion. The solution proposed in this chapter in order to determine corrosion intensity is to correlate visual deteriorations due to corrosion, in this case cracks created by expansive corrosion products, with the corrosion intensity expressed as local (LC) or generalized (GC) cross-section loss along a length of the order of the spacing between service cracks (typically an interval of 100-200 mm), as introduced in Chapter 3, above.

4.2. Determining steel reinforcement

It is important to determine the position of the reinforcements, their concrete cover and their diameter to then be able to make a correlation between deteriorations visible at the concrete surface and the corrosion rate. On-site, it is possible to use a cover meter, an instrument based on the differences in magnetic behavior between steel and concrete (non-magnetic). The Radar may also be used to perform measurements to determine the steel-reinforcement position (see [BAL 17]).

The cover meter begins by generating a low-frequency magnetic field, which induces an electrical current (Foucault's current) within the (magnetic) reinforcements located in the vicinity. After excitation has stopped, the cover-meter probe then goes on to detect the secondary field re-emitted by the magnetized reinforcements. The intensity of the magnetic signal is linked to the amount of steel and the distance between the steel and the probe, which enables the reinforcement concrete cover to be determined and offers information as to its diameter.

4.3. Corrosion diagnosis using electrochemical techniques

4.3.1. *Corrosion diagnosis using a potential mapping of the reinforcements*

Potential measurement of reinforcement steel in concrete represents the most commonly implemented technique for detecting and localizing areas of active corrosion in reinforced-concrete structures. The investigation is straightforward and inexpensive, because the measuring equipment is limited to a reference electrode, a high-impedance voltmeter and electric cables.

The general principle consists of identifying the *potential gradients* measurable at the surface of a concrete element, induced by the existence of localized anodic areas on the reinforcement layout. An electrical current flows from the corroded areas to the passive parts of the reinforcement (Figure 4.1). This current follows different lines and the equipotentials, perpendicular to these current streamlines, reach the concrete surface.

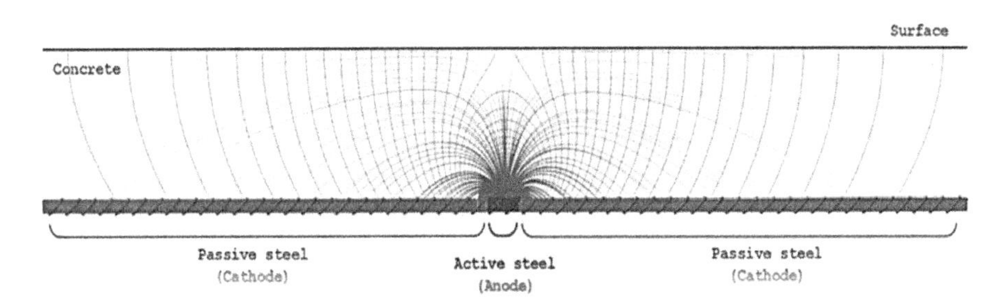

Figure 4.1. *Illustration of the potential gradient resulting from the presence of a localized anode. Results derived from numerical simulations [MAR 16]. For a color version of the figure, see www.iste.co.uk/francois/corrosion.zip*

To perform the measurement, local electrical contact needs to be established with the reinforcement layout. The reference electrode is connected to the COM (−) terminal of the voltmeter and the reinforcement layout to the (+) terminal. Next, the electrode is moved to the surface of the element (Figure 4.2).

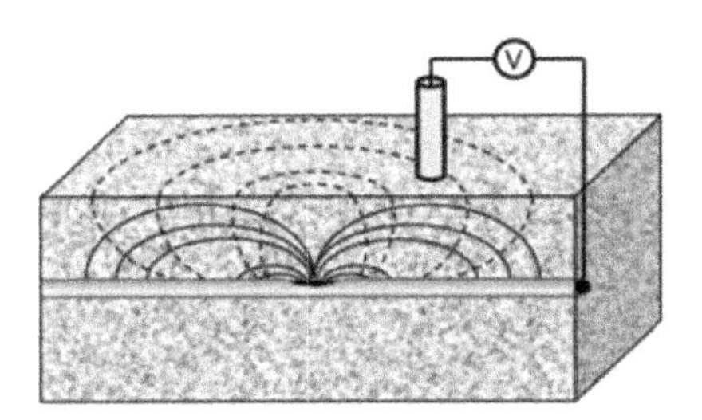

Figure 4.2. *Schematic diagram of the experimental assembly relating to the potential measurement on a reinforced-concrete structural element. The current streamlines are represented by continuous lines and the equipotentials by dotted lines. The reference electrode positioned at the concrete surface intercepts the equipotentials thanks to the voltmeter positioned between the electrode and the reinforcement. For a color version of the figure, see www.iste.co.uk/francois/corrosion.zip*

It ought to be noted that it is necessary to ensure correct electrical contact between electrode and concrete at each measurement point. A damp sponge is often used to this end. Surface potential measurement is carried out according to a regular grid in order to be able to plot a potential map enabling simple identification of the areas of active corrosion (Figure 4.3).

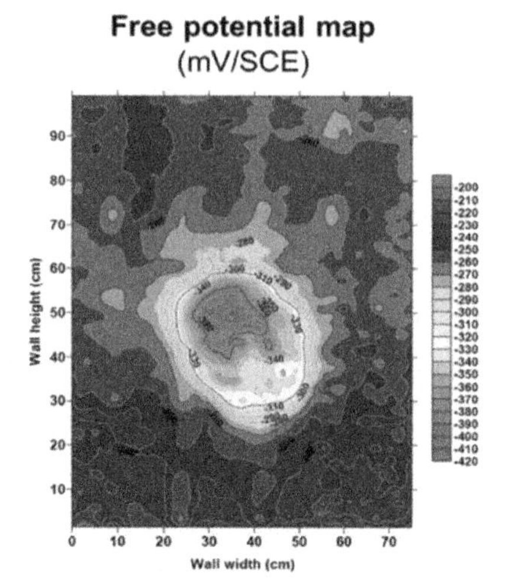

Figure 4.3. *Example of a potential map collected at the surface of a concrete wall [SAS 16] – The active corrosion site is clearly identified by the significant drop in potential at the center of the image. For a color version of the figure, see www.iste.co.uk/francois/corrosion.zip*

The regulatory context associated with the potential measurement is based on an American standard, ASTM C-876 (revised 2009) [AST 09], and a technical recommendation from TC 154-EMC (2003) [ELS 03] by a RILEM committee of European experts. The localized nature of corrosion is explicitly recognized there, with it being pointed out that it is the most common type of corrosion in reinforced concrete. The RILEM recommendation, drawn up in 2003, defines the *in situ* measurement experimental protocol, as well as the result interpretation modes. The major requirement set out in the recommendation involves the detection of potential gradients at the surface, as this is the only information to reveal the existence of one or more areas of active corrosion.

Prior to the revision applied in 2009, the ASTM standard called on the absolute potential measurements to be interpreted in terms of corrosion risk. This interpretation mode has been widely used, notably with a view to detecting corrosion initiation. The value E_{corr} = −200 mV/CSE was then considered to be an initiation criterion. In the revised version of the standard (2009), this interpretation protocol was reviewed and, like the RILEM recommendation, it now recommends examining the *potential gradients* in order to identify the areas of active corrosion.

The fact that potential gradient is considered to be relevant information in identifying the areas of active corrosion means that the galvanic nature of corrosion systems in reinforced concrete is explicitly acknowledged. Indeed, in the case of a uniform corrosion system, the potential field is theoretically uniform within the volume and, consequently, at the surface of the element. In practice, a uniform potential field can only be observed in the case of a passive steel-reinforcement layout. When corrosion is initiated locally, the potential field is modified over a significant distance with respect to the initiation area, it loses its uniformity and reflects the respective positions of the active and passive sites.

It should be noted that interpretation in terms of potential gradients makes it possible to dispense with a direct connection on the reinforcement layout: it is thus possible to conduct a measurement with a reference *two-electrode system* (one stationary and one mobile, or two mobile), as demonstrated by François *et al.* in 1994 [FRA 94c].

It should also be noted that interpreting in terms of potential gradients also makes it possible to dispense with the existence of electrical continuity on the reinforcement layout, and therefore with its verification, as demonstrated by Sassine *et al.* [SAS 17].

NOTE.– The measurement of the reinforcement potential on the concrete surface offers an idea, in qualitative terms, of the presence of active localized corrosion, identified by potential gradients. No quantitative information is possible however, regarding the extent of the corroded areas, the corrosion intensity in terms of cross-section loss, or its evolution in terms of kinetics.

4.3.2. *Corrosion diagnosis based on a linear polarization resistance measurement*

The measurement of the linear polarization resistance of steel in concrete is frequently used with a view to assessing the steel dissolution kinetics within a proven corrosion area. This method, like the conventional potential measurement method at a single electrode, requires a connection to the reinforcement. The measurement duration is longer as the method consists of injecting a polarization current in order to shift the electrochemical equilibrium of the corrosion system characterized by its free potential, E_{corr}, at each measurement point.

A RILEM recommendation from the 154-EMC technical committee established, in 2004 [AND 04b], the sole regulatory framework relating to this measurement technique within the field of reinforced concrete. The recommendation defines the protocol for implementing *in situ* measurement and the interpretation method dedicated to assessing corrosion current. The RILEM recommendation advises the use of certain instruments incorporating a polarizing-current confinement device. Indeed, the appearance of confinement devices in measurement systems resulted from the recurrent questions posed regarding the steel area effectively concerned by the measurement, often referred to by the term *Polarized Surface*.

In order to localize the measurement in a controlled area, devices available commercially therefore incorporate an additional electrode, known as a "guard electrode". Owing to the annular shape adopted in devices dedicated to measurements in reinforced concrete, these guard electrodes are often known as "guard-rings".

Figure 4.4 presents and comments on an illustration taken from the 2004 RILEM recommendation [MAR 16]. The system represented (Gecor®) is characterized by an annular counter-electrode and a reference electrode, placed at the center of the measurement probe. A guard ring encircles the counter-electrode. Lastly, the presence of two additional reference electrodes, S_1 and S_2, can be noted. The mode of operation of a guard ring consists of injecting an additional current, I_{gr}, intended to confine the counter-electrode current, I_{ce}, within a clearly defined area. The current value, I_{gr}, is controlled in order to keep the initial potential difference measured between probes S_1 and S_2 constant. This protocol aims to cancel the horizontal potential gradient induced by injecting the counter-electrode current and as such focus the polarizing current perpendicularly to the element's surface. To do so, the injected currents, I_{ce} and I_{gr}, must present the same sign.

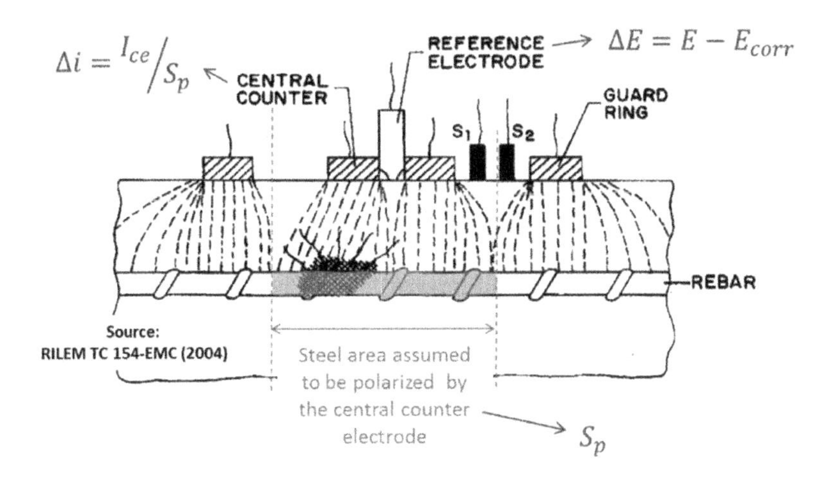

Figure 4.4. *Illustration of a measurement with confinement device [MAR 16]. For a color version of the figure, see www.iste.co.uk/francois/corrosion.zip*

According to this protocol, the counter-electrode current, I_{ce}, is assumed to polarize a steel surface situated in the area indicated in Figure 4.4. This assumed-to-be-polarized surface, designated S_p below, constitutes a sort of measurement constant linked to the size of the probe and to the diameter of the reinforcement located below it. Given the dimensions of the Gecor® probe, the surface considered to be polarized can thus vary from around 10 to 150 cm², according to the number and diameter of the reinforcements located beneath the probe.

The apparent polarization resistance measurement (R_p in $\Omega.m^2$) is then determined in a conventional manner (equation [4.1]):

$$R_p = \frac{\Delta E}{\Delta I} = \frac{E - E_{corr}}{I_{ce} / S_p} \qquad [4.1]$$

The measurements carried out by means of commonly-used devices (Gecor®, Galvapulse®) are based on a galvanostatic action. The constant polarizing current, I_{CE}, is impressed and the temporal response, $\Delta E(t)$, is measured over a time period chosen by the user. An ohmic drop correction is then applied, often based on Randles' equivalent electrical circuit, with a view to erasing the ohmic effects generated by the electrical resistivity of the cover concrete. The polarization resistance obtained in this manner is converted into corrosion current density, i_{corr}, using the Stern-Geary equation (equation [4.2]) [STE 57].

$$i_{corr} = \frac{B}{R_p} \qquad [4.2]$$

Based on empirical results, the RILEM protocol defines two possible values for the constant B, depending on the electrochemical state of the steel, 26 mV for an active steel, 52 mV for a passive steel.

Lastly, the measurements are interpreted with respect to the corrosion kinetics scale presented in Table 4.1 and drawn up empirically using the Gecor® device.

i_{corr} (µA/cm²)	V_{corr} (mm/year)	Corrosion level
≤ 0,1	≤ 0,001	Negligible
0,1 – 0,5	0,001 – 0,005	Low
0,5 – 1	0,005 – 0,01	Moderate
> 1	> 0,01	High

Table 4.1. *Corrosion kinetics measurement interpretation table, according to RILEM TC-154-EMC*

The linear polarization resistance measurement is a solid example of the unsuitability of conventional electrochemical techniques for the phenomenological reality of steel corrosion in concrete. Indeed, this technique is based on a double uniformity assumption: uniformity of the reinforcement corrosion state and uniformity of the electrochemical action imposed for metrological reasons. Thus, the physical assumptions relating to the interpretation of R_p measurements contradict the assumptions formulated within the framework of potential measurements. The scientific community thus agrees on the fact that this measurement is not controlled at present. As an example, different measurement devices tested on the same corrosion system generate significantly different results in terms of kinetics [POU 05].

Based on the above brief description of the measurement of R_p as recommended by the RILEM TC 154-EMC expert committee, it is possible to reconsider a certain number of points regarding the physical assumptions adopted. First of all, Figure 4.4 explicitly represents an example of localized corrosion (an explicitly drawn localized anodic area). Thus, the first major criticism concerns the lack of consideration as to the natural galvanic current exchanged between the active site and passive sites, a current of the same order of magnitude as the impressed current, I_{ce}. This representation therefore does not restore interaction between the galvanic corrosion current, I_m, and the impressed current, I_{ce}, and therefore offers at the very least a truncated view of measurement physics.

The distribution of currents I_{ce} and I_{gr} represented in Figure 4.4 corresponds, in reality, to the case of a bar in uniform corrosion state (uniform electrochemical state). Thus, these schematic diagrams tend to explain a measurement carried out on a localized corrosion system by considering it like a uniform corrosion system. It would appear that these schematic diagrams reflect an assumed notion of the measuring device, and not a notion bolstered by a theory-based calculation.

The measurement, using a Gecor®-type device, on a galvanic corrosion system has been studied by means of a series of numerical simulations [MAR 16]. The distributions of currents I_{ce} and I_{gr} obtained in Figure 4.5 for an anodic polarization are notably different from those of Figure 4.4. The polarization current and guard current behave in exactly the same way: the guard ring therefore also contributes fully to the polarization of the active area, whereas I_{gr} is overlooked when calculating the polarization resistance, R_p.

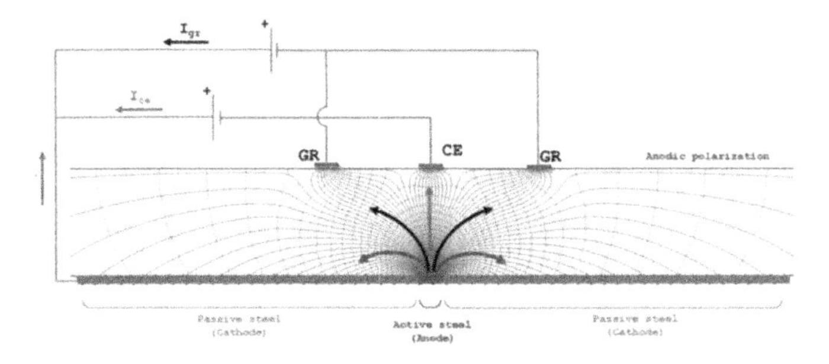

Figure 4.5. *Numerical simulation of a GECOR® -type measurement in anodic polarization [MAR 16]. For a color version of the figure, see www.iste.co.uk/francois/corrosion.zip*

Another fundamental point focuses on the applicability of the Stern-Geary equation in the case of localized corrosion. The 2004 RILEM recommendation calls on the use of the Stern-Geary equation to calculate corrosion current. The localized nature of corrosion through the different schematic diagrams proposed is explicit, but the measured data is interpreted based on the assumption of uniform corrosion. Indeed, the conventional concept of linear polarization resistance is defined within the strict framework of uniform corrosion, where active and passive sites are spatially mixed up, such that the system equilibrium is characterized by a uniform potential, E_{corr}, within the entire concrete volume.

Lastly, to offer a qualitative summary of the response of a galvanic system to an electrical action, Figure 4.6 presents the distribution of the current injected by a counter-electrode in a case of localized corrosion [LAU 16]. These results highlight the difference in the system's behavior according to the polarization direction (anodic or cathodic).

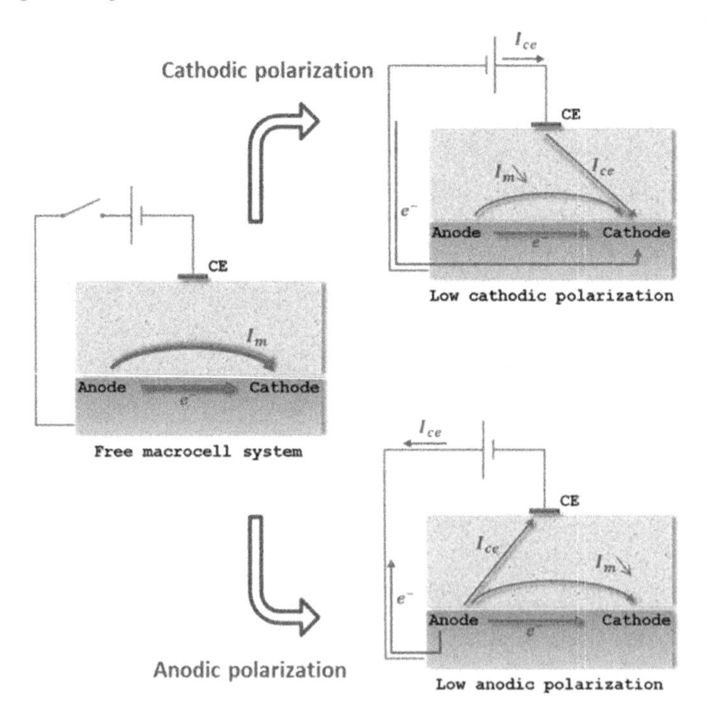

Figure 4.6. *Schematic behavior of the polarization current of a galvanic system according to the anodic and cathodic directions [LAU 16]. For a color version of the figure, see www.iste.co.uk/francois/corrosion.zip*

NOTE.– The linear polarization resistance measurement, which does not take into account the localized nature of reinforcement corrosion in reinforced concrete, does not enable the corrosion kinetics to be quantified. Nor does this method enable the active areas of corrosion to be localized.

Given the absence at present of reliable tools for electrochemical diagnosis of corrosion in a reinforced concrete structure, we choose in this book to rely on the observation of visible corrosion consequences: corrosion-induced cracks.

4.4. Quantification of corrosion rate by recording the cracks created by reinforcement corrosion

One possibility for quantifying the reinforcement corrosion rate and to establish a correlation between the visible consequences of corrosion is to focus on the cracks induced at the concrete surface and the reinforcement cross-section loss. To make this correlation, it is necessary to determine for which corrosion rate the creation of the first corrosion cracks is obtained, because the increase in crack opening will be correlated with the increase in corrosion beyond the value initiating the first corrosion cracks.

4.4.1. *Rate of local corrosion triggering corrosion-induced cracks*

Following the start of active corrosion, there is a period of propagation prior to the occurrence of the first corrosion cracks, which are due to the expansive nature of the corrosion products (Figure 4.7), thus exerting a pressure on the cover concrete.

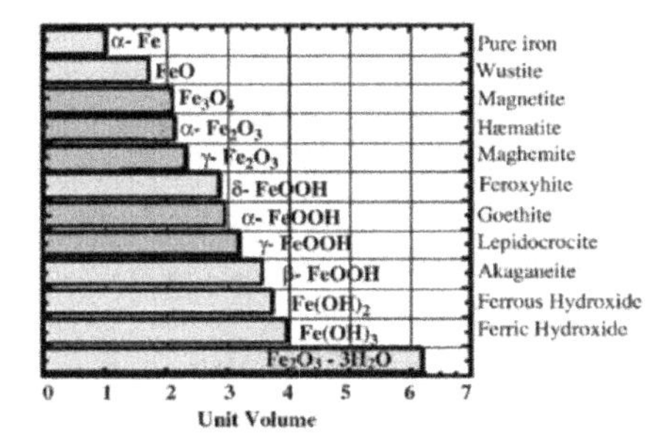

Figure 4.7. *Corrosion products and their expansion with respect to the initial iron volume, according to Marcotte [MAR 01]*

Different approaches exist to predict the occurrence of the first corrosion cracks. Thus, experimental approaches are found, based on the use of accelerated tests under electrical field [AND 93, MUL 11, VU 05, ELM 07], which are not representative of actual corrosion. Calculation models are also to be found that use in-plane stress [LUN 05], based on a uniform expansion of the reinforcement surface. This approach does not take into account the localized nature of chloride-induced corrosion. The approach used in this chapter is that implemented by Vidal *et al.* [VID 04] via the autopsy of beams corroded naturally for several years. This approach involves considering that the pressure exerted by the corrosion products relates to the local cross-section loss. The model is empirical and is set according to the local corrosion measurements, without the presence of corrosion cracks, which limit the lower cross-section loss value, and the local cross-section loss measurement for tiny corrosion cracks, which limits the upper cross-section loss value. *The influence of very localized corrosion (pitting) and more homogeneous corrosion (carbonation case) is taken into account by the geometric pitting factor,* p_{fg}. Equation [4.3], adapted from that proposed by Vidal *et al.* [VID 04] thus enables the local cross-section loss initiating cracking to be determined. The reinforcement diameter and the distance between the reinforcement and the concrete form part of this calculation, but not the tensile concrete strength.

$$\Delta A_{s0} = A_s\left[1-\left[1-\frac{2pf_g}{\phi_0}\left(7.53+9.32\frac{c}{\phi_0}\right)10^{-3}\right]^2\right] \qquad [4.3]$$

In equation [4.3], ΔA_{s0} is the local cross-section loss leading to the occurrence of corrosion-induced cracks in mm^2, A_s is the cross-section of the reinforcement in question (mm^2), ϕ_0 is the nominal reinforcement diameter (mm), c is the distance between the reinforcement and the outer concrete surface (mm) and pf_g is the geometric pitting factor, taken as equal to 4 for chloride-induced corrosion and 1 for carbonation-induced corrosion.

As an example, let us address the calculation based on equation [4.3] with another beam, Bs03, of which the transverse dimensions are presented in Figure 4.8. This beam has been kept under load for 19 months in an environment with 35 g/l salt-spray spraying-and-drying cycles (two days of spraying and 12 days of drying) [YU 15a]. Core samples were taken parallel to the tensile longitudinal reinforcements and observed by X-ray Tomography at BAM in Berlin.

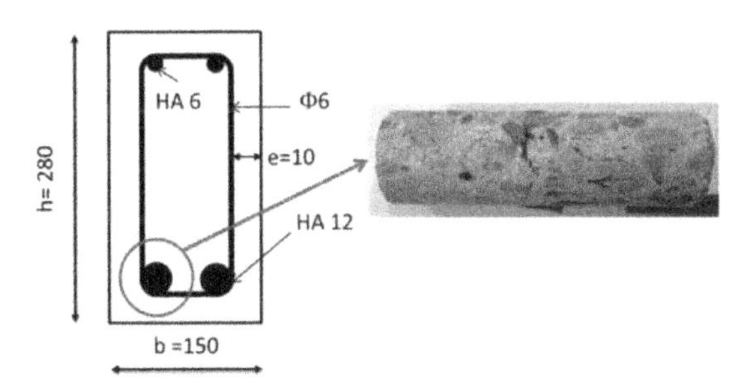

Figure 4.8. *Transverse cross-section of Beam Bs03 and view of a core sample taken along the tensile longitudinal reinforcements. For a color version of the figure, see www.iste.co.uk/francois/corrosion.zip*

Figure 4.9 shows two cross-sections of a core sample with a view of a chloride-induced corrosion pit presenting, in one case, (a) a corrosion crack created by the pit and, in the other case, (b) no pit-induced corrosion cracking.

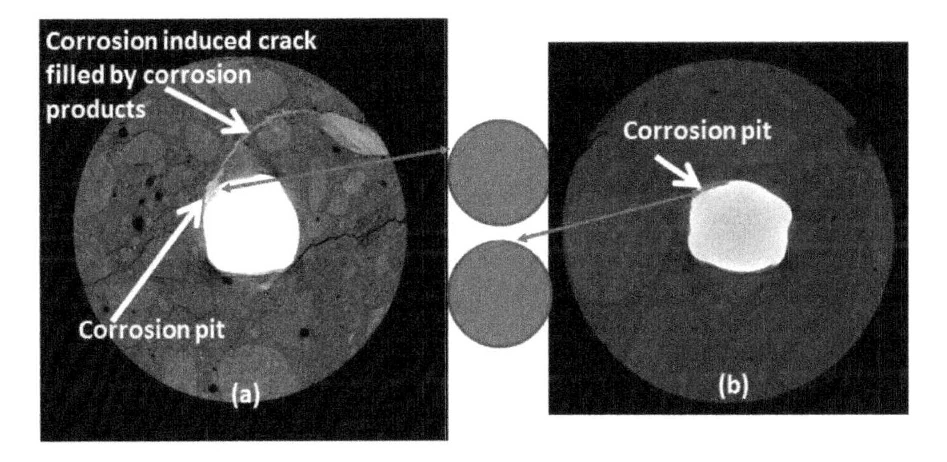

Figure 4.9. *Transverse cross-sections of the concrete area surrounding a tensile reinforcement, 12 mm in diameter, of beam Bs03 with two corrosion pits: one that has generated a corrosion crack, and one that has not generated a corrosion crack (X-ray Tomography (BAM Berlin)). For a color version of the figure, see www.iste.co.uk/francois/corrosion.zip*

The steel cross-section loss calculation corresponding to the two corrosion pits seen in Figure 4.9 is produced graphically, and the values are indicated in Table 4.2: it can be noted that the cross-section initiating cracking, proposed by Vidal *et al.*, is well enclosed by the two values corresponding respectively to the lack and presence of a corrosion crack (Figure 4.9).

cross-section loss initiating cracking	ΔA_{s0} mm^2	ΔA_s mm^2
Vidal *et al.*	2.99	
cross-section (a)	3.4	
cross-section (b)	N/A	2.5

Table 4.2. *Comparison between cross-section loss due to corrosion initiating cracking calculated by the model by Vidal* et al.*, and the cross-section losses corresponding to the absence or presence of a corrosion crack, calculated from Figure 4.6*

The value of the cross-section loss initiating corrosion is low and in the example presented in Figure 4.9, represents only 2.6% of the reinforcement cross-section. Such orders of magnitude do not present any consequences for the structural behavior of reinforced-concrete structures.

4.4.2. *Recording corrosion crack openings*

Beyond a minimum corrosion rate, defined in the previous section, concrete reinforcement corrosion leads to the occurrence of cracks (referred to as "corrosion cracks"), coinciding with the reinforcements. Indeed, unlike service cracks, which are linked to mechanical stress (reflecting the tensile weakness of concrete), and which are therefore generally perpendicular to the main reinforcements (see Figure 4.10), corrosion cracks are linked to the pressure induced by expansive corrosion products and are therefore parallel to the reinforcements (see Figure 4.11).

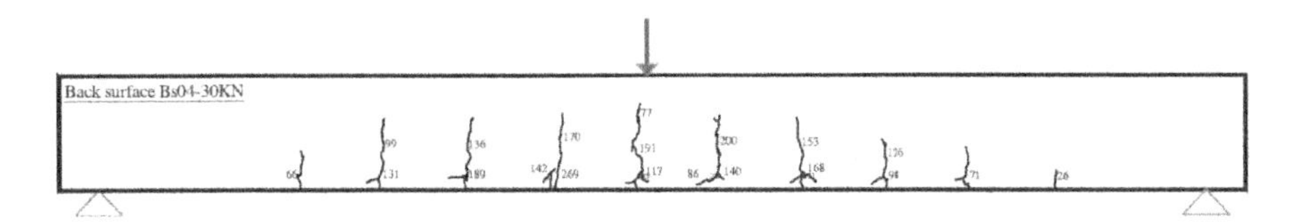

Figure 4.10. *Recording of the transverse service cracks on a reinforced-concrete beam, Bs04 ([YU 15b]). For a color version of the figure, see www.iste.co.uk/francois/corrosion.zip*

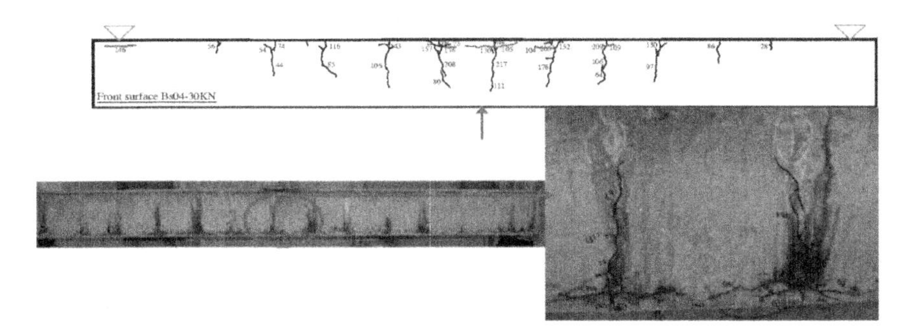

Figure 4.11. *View of the longitudinal corrosion cracks and transverse service cracks on a reinforced-concrete beam, Bs04 (photo credit: Raoul François). For a color version of the figure, see www.iste.co.uk/francois/corrosion.zip*

Using binoculars, it is possible, although tedious, to record the corrosion crack openings on all of the accessible faces of a beam or, more generally, of a reinforced-concrete structure. As an example, Figure 4.12 shows the recording of the corrosion crack openings on the lateral faces and tensile face of a reinforced-concrete beam, performed by Yu *et al.* [YU 15b].

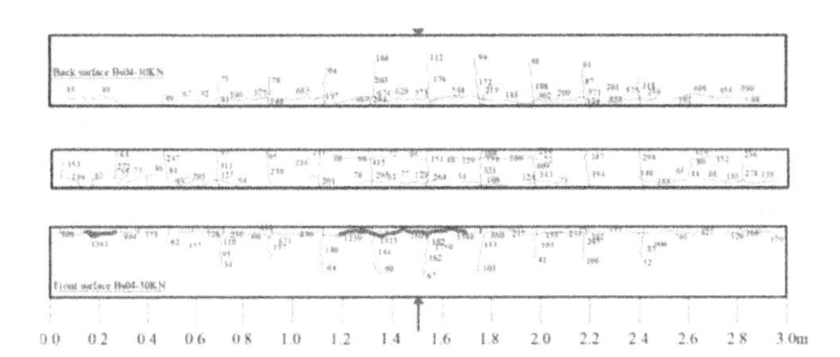

Figure 4.12. *Recording of the openings of longitudinal corrosion cracks and transverse service cracks on a reinforced-concrete beam, Bs04, according to Yu* et al. *[YU 15b]. For a color version of the figure, see www.iste.co.uk/francois/corrosion.zip*

The recording of the corrosion crack openings is a signature of the state of deterioration of a concrete structure and can therefore be correlated with the corrosion intensity. This recording can provide several pieces of information. Firstly, it can offer an idea of the duration of the corrosion process, secondly, an idea of the maximum local corrosion intensity (identified by the local loss, LC), and lastly, an idea of the average corrosion intensity across the distance between two service cracks (identified by the generalized loss, GC). The corrosion process duration is useful in order to know whether we are dealing with essentially localized corrosion (during the first years), or whether we have gradually superposed a more global corrosion onto the localized corrosion, due to the change in environment around the reinforcements, linked to the creation of corrosion cracks. In the first case, local loss will be the prevalent parameter for the structure's corrosion diagnosis, and in the second case, local loss (LC) and average loss (GC) will both be very important.

Figure 4.13 illustrates the difference and non-correlation between the local loss, LC, which represents the maximum possible loss on a bar coupon (localized corrosion) and the average loss, GC, which represents the amplitude of the generalized corrosion and not the local loss divided by the length of the reinforcement coupon.

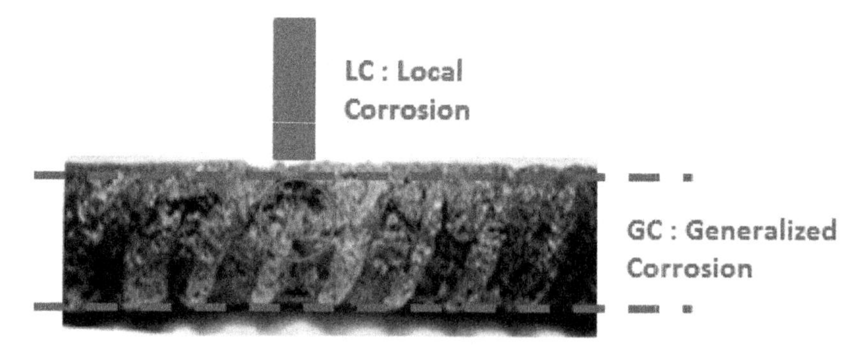

Figure 4.13. *Illustration of the difference between local corrosion loss, LC, due to localized corrosion, and average loss, GC, due to generalized corrosion. For a color version of the figure, see www.iste.co.uk/francois/corrosion.zip*

The ratio of local loss (LC) to generalized loss (GC) corresponds to pitting factor pf_{lc}, introduced in the previous chapter. It is impossible to quantify in principle.

4.4.3. *Correlation between corrosion crack openings and maximum LOCAL corrosion (localized corrosion, LC)*

In order to predict the mechanical behavior of corroded structures, it is necessary to have an estimation of the rate of local corrosion in the structure. This point will be highlighted in Chapters 5 and 6, examining the consequences of corrosion on mechanical behavior and the recalculation of bearing capacity.

Prediction of corrosion-induced cross-section loss based on corrosion crack opening only concerns the reinforcements that are closest to the surface, and for which we allocate one or more cracks assumed to result from the pressures induced by the corrosion products developed on the reinforcement in question. It is therefore essential to have accurately identified the position of the reinforcements and their diameter (see section 4.2). The crack opening calculated for a given reinforcement bar, based on the visual measurement of the corrosion crack openings, is known as the equivalent opening, w_{eq}. This concept is illustrated in Figure 4.14. If two cracks of openings w_1 and w_2, are attributed to a reinforcement, the equivalent opening, w_{eq}, is equal to the sum of the crack openings, w_1+w_2.

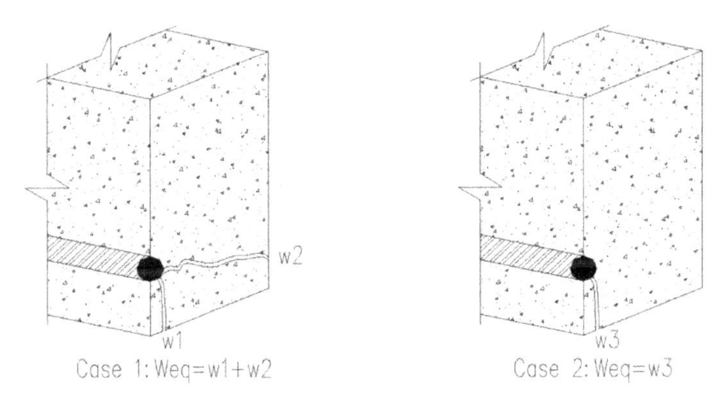

Figure 4.14. *Example of the determination of two equivalent crack openings, according to Khan* et al. *[KHA 14]*

The correlation between the equivalent crack opening and local corrosion rate is based on the experimental observation that the crack opening increases with the corrosion duration and therefore as a function of the amount of corrosion products formed (equation [4.4]).

$$\Delta A_s = \frac{w_{éq}}{K} + \Delta A_{s0} \qquad [4.4]$$

Whereby w_{eq} is the equivalent crack opening in mm, ΔA_s is the local cross-section loss in mm^2, ΔA_{s0} is the cross-section loss initiating corrosion cracking in mm2 (defined in section 4.4.1 above) and K=0.0575 a proportionality coefficient.

The proportionality coefficient, K=0.0575, was drawn up by Vidal *et al.* [VID 04] based on the experimental results obtained on two corroded beams in saline environment, B1CL1 and A1CL1, for 14 and 17 years, respectively. A confirmation of the validation of this model was proposed by Khan *et al.* [KHA 14] on another beam, A2CL3, aged for 26 years (see Figure 4.15).

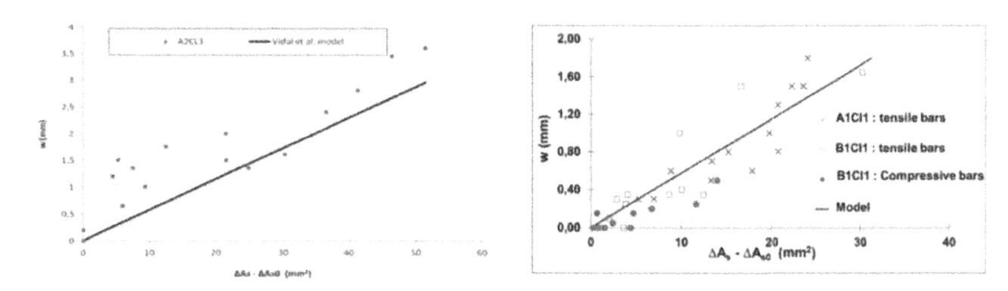

Figure 4.15. *Empirical model for prediction of local cross-section loss by Vidal* et al. *[VID 04] and validation by Khan* et al. *[KHA 14]. For a color version of the figure, see www.iste.co.uk/francois/corrosion.zip*

Given that there can be internal corrosion cracks that are not visible at the surface (see Figure 4.16), this model by Vidal *et al.* can underestimate the corrosion in areas where there is no visible corrosion crack on the surface.

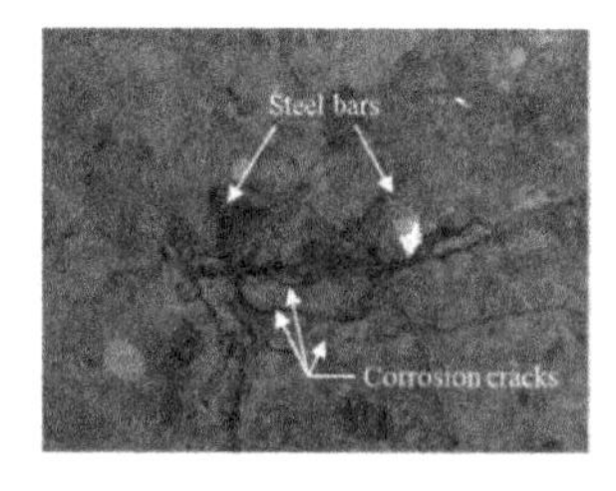

Figure 4.16. *Example of presence of internal corrosion cracks that are not taken into account in the equivalent opening calculation, according to Khan* et al. *[KHA 14]. For a color version of the figure, see www.iste.co.uk/francois/corrosion.zip*

The localized corrosion (LC) calculation model must be understood to be the upper bound of possible corrosion pits. With the corrosion pits being of reduced length, it is not possible to deduce an average corrosion (generalized across a reinforcement length of 100 to 200 mm) using this model. A prediction of this generalized corrosion (GC) forms the subject of section 4.4.4. that follows.

4.4.4. *Correlation between corrosion crack openings and generalized corrosion (GC) across a length neighboring the spacing between the service cracks, or the shear stirrups in the absence of service cracks*

During the oxidation process, the reinforcements plumb over the corrosion cracks see their conditions evolve and become the site of generalized corrosion. This phenomenon is presented as a schematic diagram in Figure 4.17, taken from the work of Zhang *et al.* [ZHA 10].

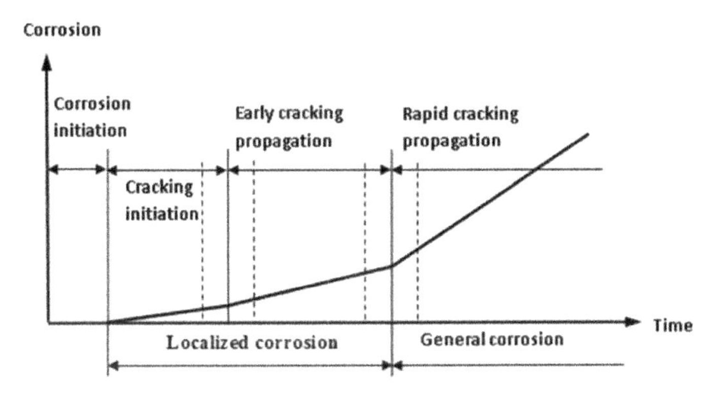

Figure 4.17. *Evolution of the corrosion process, first localized, then generalized, according to Zhang* et al. *[ZHA 10]*

In the presence of generalized corrosion that is superposed onto the localized corrosion, it becomes important to be able to plan for an average cross-section loss ΔA_{sm} having an influence on the mechanical behavior. The local loss remains necessary for the behavior at failure, but an average loss over the distance between service cracks (under bending, for example) will be necessary to plan for the evolution of the deformations and deflections. In the absence of service cracks, the distance retained will be the spacing between shear stirrups. The phenomenology of local loss, ΔA_s and average cross-section loss, ΔA_{sm}, due to corrosion can be seen in Figure 4.18.

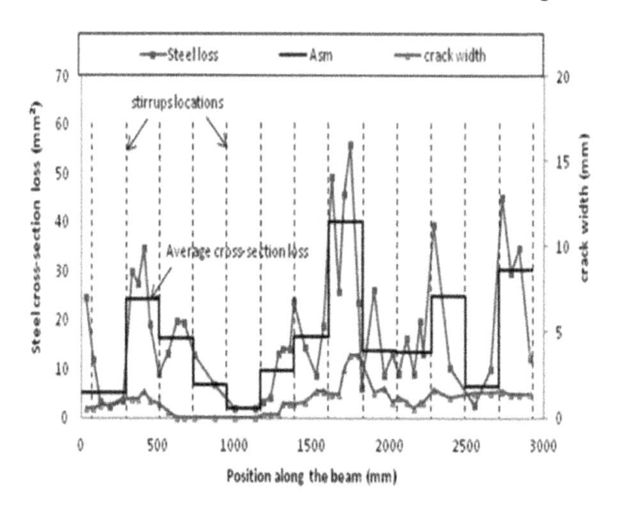

Figure 4.18. *Local corrosion (LC) in blue and average corrosion (GC) in black on a reinforced-concrete beam, A2CL3, according to Khan* et al. *[KHA 14]. For a color version of the figure, see www.iste.co.uk/francois/corrosion.zip*

The process of generalized corrosion (GC) appears to be sensitive to the concrete cover and to the diameter of the reinforcements [ALH 11, KHA 14], unlike the localized corrosion (LC) process, as defined by Vidal *et al.* [VID 04]. As a consequence, Khan *et al.* [KHA 14] propose a modification to the initial model by Zhang *et al.* [ZHA 10] for the calculation of the average cross-section loss, taking into account the ratio of cover to diameter of the reinforcement (equation [4.5]).

$$\Delta A_{sm} = 5.22\frac{c}{\varphi}(w_{eq} - 0.164) \quad [4.5]$$

Whereby w_{eq} is the equivalent crack opening defined in Figure 4.14, in mm (> 0.164 mm, otherwise there is only localized corrosion), c is the reinforcement concrete cover, ϕ is the reinforcement diameter and ΔA_{sm} is the average loss (GC) of cross-section in mm^2.

4.4.5. *Example of corrosion diagnosis based on a visual examination of the concrete surface*

In this section, we propose an application of the calculation of local (LC) and average (GC) corrosion rate based on the recording of corrosion cracks, made in section 4.3.1 and shown in Figure 4.12, of beam Bs04. Figure 4.19 presents the notations used to describe the different faces of the beam: *f* (front) for a lateral face, *b* (back) for the opposite face and *t* (tensile) for the tensile lower surface. The corrosion cracks located on the tensile face are attributed either to the tensile reinforcement on side *f* (front) or to the tensile reinforcement on side *b* (back), according to their location. The diameter, Φ, of the tensile reinforcements is 12 mm and parameter c = 16 mm corresponds to the distance between the reinforcement and the beam surface.

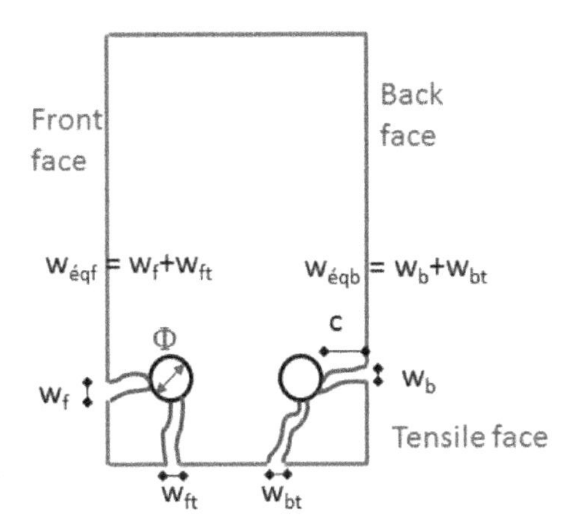

Figure 4.19. *Cross-section of beam Bs04 and view of the geometric parameters and the crack recording. For a color version of the figure, see www.iste.co.uk/francois/corrosion.zip*

The equivalent crack opening is calculated with a pitch of 100 mm, each time considering the maximum crack opening recorded on the front and back lateral faces, as well as on the tensile surface (see Figure 4.19).

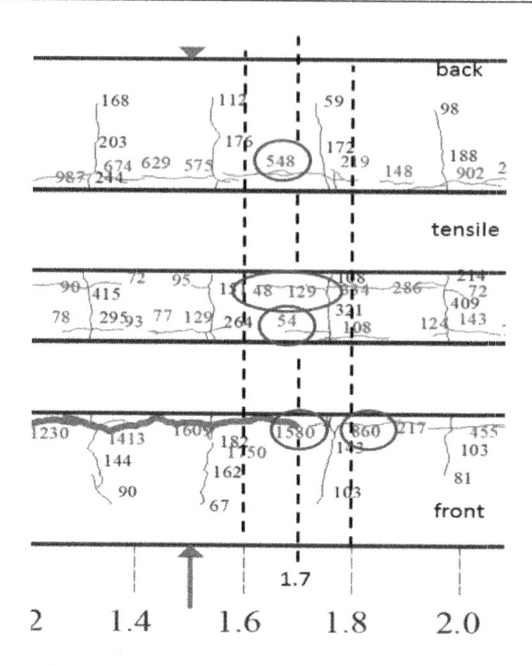

Figure 4.20. *Recording of the crack openings by surface of beam Bs04 in 10^{-2} mm. For a color version of the figure, see www.iste.co.uk/francois/corrosion.zip*

Figure 4.20 offers a close-up view of the cracking map of beam Bs04, for better understanding of the calculation process at the interval 1.2 m - 2.0 m. Thus, for the interval 1,600-1,700 mm, the following is recorded: wb = 0.548 mm; wbt = 0.048 mm, wf = 1.58 mm; wft = 0 mm and for the interval 1,700-1,800 mm, the following is recorded: wb = 0.548 mm; wbt = 0.0129 mm, wf = 0.86 mm; wft = 0.054 mm. These values are reported in Table 4.3 for each of the 100-mm intervals along the beam 3 m in length. The equivalent crack opening values are then calculated by equations [4.6] and [4.7]:

$$w_{éqf} = w_f + w_{ft} \quad [4.6]$$

$$w_{éqb} = w_b + w_{bt} \quad [4.7]$$

Interval (mm)	w_f (mm)	w_{ft} (mm)	w_{eqf} (mm)	w_b (mm)	w_{bt} (mm)	w_{eqb} (mm)
0–100	0.509	0	0.509	0.043	0	0.043
100–200	1.361	0.119	1.48	0.043	0	0.043
200–300	1.361	0.062	1.423	0.049	0	0.049
300–400	0.944	0.073	1.017	0	0	0
400–500	0.151	0.036	0.187	0	0	0
500–600	0.457	0.083	0.54	0.067	0	0.067
600–700	0.328	0.295	0.623	0.092	0	0.092
700–800	0.236	0.127	0.363	0.092	0	0.092
800–900	0.111	0	0.111	0.19	0	0.19
900–1,000	0.671	0	0.671	0.375	0	0.375
1,000–1,100	0.836	0	0.836	0.685	0	0.685
1,100–1,200	1.23	0	1.23	0.685	0.216	0.901
1,200–1,300	1.413	0	1.413	0.987	0.086	1.073
1,300–1,400	1.609	0.078	1.687	0.987	0.09	1.077
1,400–1,500	1.75	0.077	1.827	0.629	0.072	0.701
1,500–1,600	1.75	0.129	1.879	0.575	0.095	0.67
1,600–1,700	1.58	0	1.58	0.548	0.048	0.596
1,700–1,800	0.86	0.054	0.914	0.548	0.129	0.677
1,800–1,900	0.86	0.108	0.968	0.219	0.334	0.553
1,900–2,000	0.217	0.124	0.341	0.148	0.286	0.434
2,000–2,100	0.455	0.143	0.598	0.902	0.072	0.974
2,100–2,200	0.233	0	0.233	0.209	0	0.209
2,200–2,300	0.382	0	0.382	0.635	0	0.635
2,300–2,400	0.211	0	0.211	0.575	0	0.575
2,400–2,500	0.906	0.188	1.094	0.239	0	0.239
2,500–2,600	0.74	0.065	0.805	0.182	0	0.182
2,600–2,700	0.427	0.084	0.511	0.608	0.124	0.732
2,700–2,800	0.12	0.181	0.301	0.454	0.152	0.606
2,800–2,900	0.166	0.138	0.304	0.39	0	0.39
2,900–3,000	0.179	0.138	0.317	0	0	0

Table 4.3. *Recording of the crack openings on each face of beam Bs04 and calculation of the equivalent crack opening in correspondence with the two tensile reinforcements*

Interval (mm)	w_{eqf} (mm)	w_{eqb} (mm)	ΔA_{sf} mm^2	ΔA_{sb} mm^2	ΔA_s mm^2	C%
0–100	0.509	0.043	11.8	3.7	15.5	6.9%
100–200	1.48	0.043	28.7	3.7	32.4	14.3%
200–300	1.423	0.049	27.7	3.8	31.5	13.9%
300–400	1.017	0	20.7	3	23.7	10.5%
400–500	0.187	0	6.2	3	9.2	4.1%
500–600	0.54	0.067	12.4	4.2	16.6	7.3%
600–700	0.623	0.092	13.8	4.6	18.4	8.1%
700–800	0.363	0.092	9.3	4.6	13.9	6.1%
800–900	0.111	0.19	4.9	6.3	11.2	5.0%
900–1,000	0.671	0.375	14.7	9.5	24.2	10.7%
1,000–1,100	0.836	0.685	17.5	14.9	32.4	14.3%
1,100–1,200	1.23	0.901	24.4	18.7	43.1	19.1%
1,200–1,300	1.413	1.073	27.6	21.7	49.3	21.8%
1,300–1,400	1.687	1.077	32.3	21.7	54	23.9%
1,400–1,500	1.827	0.701	34.8	15.2	50	22.1%
1,500–1,600	1.879	0.67	35.7	14.6	50.3	22.2%
1,600–1,700	1.58	0.596	30.5	13.4	43.9	19.4%
1,700–1,800	0.914	0.677	18.9	14.8	33.7	14.9%
1,800–1,900	0.968	0.553	19.8	12.6	32.4	14.3%
1,900–2,000	0.341	0.434	8.9	10.5	19.4	8.6%
2,000–2,100	0.598	0.974	13.4	19.9	33.3	14.7%
2,100–2,200	0.233	0.209	7	6.6	13.6	6.0%
2,200–2,300	0.382	0.635	9.6	14	23.6	10.4%
2,300–2,400	0.211	0.575	6.7	13	19.7	8.7%
2,400–2,500	1.094	0.239	22	7.1	29.1	12.9%
2,500–2,600	0.805	0.182	17	6.2	23.2	10.3%
2,600–2,700	0.511	0.732	11.9	15.7	27.6	12.2%
2,700–2,800	0.301	0.606	8.2	13.5	21.7	9.6%
2,800–2,900	0.304	0.39	8.3	9.8	18.1	8.0%
2,900–3,000	0.317	0	8.5	3	11.5	5.1%

Table 4.4. *Calculation of the local cross-section loss, ΔA_S, for each of the tensile bars, then of the total value for a beam cross-section*

Interval (mm)	w_{eqf} (mm)	w_{eqb} (mm)	ΔA_{smf} mm^2	ΔA_{smb} mm^2	ΔA_{sm} mm^2	$C_m\%$
0–200	1.48	0.043	9.2	0	9.2	**4.1%**
200–400	1.423	0.049	8.8	0	8.8	**3.9%**
400–600	0.54	0.067	2.6	0	2.6	**1.1%**
600–800	0.623	0.092	3.2	0	3.2	**1.4%**
800–1,000	0.111	0.19	0	0.2	0.2	**0.1%**
1,000–1,200	0.836	0.685	4.7	3.6	8.3	**3.7%**
1,200–1,400	1.687	1.077	10.6	6.4	17	**7.5%**
1,400–1,600	1.879	0.67	11.9	3.5	15.4	**6.8%**
1,600–1,800	1.58	0.596	9.9	3	12.9	**5.7%**
1,800–2,000	0.968	0.553	5.6	2.7	8.3	**3.7%**
2,000–2,200	0.598	0.974	3	5.6	8.6	**3.8%**
2,200–2,400	0.382	0.635	1.5	3.3	4.8	**2.1%**
2,400–2,600	1.094	0.239	6.5	0.5	7	**3.1%**
2,600–2,800	0.511	0.732	2.4	4	6.4	**2.8%**
2,800–3,000	0.304	0.39	1	1.6	2.6	**1.1%**

Table 4.5. *Calculation of the average cross-section loss, ΔA_{sm}, for each of the tensile bars, then of the total value for a beam cross-section*

Based on equivalent crack openings, the local cross-section loss can be calculated with the model by Vidal *et al.* (equation [4.4]) for each of the two tensile bars of beam Bs04, then the total loss for the two bars in a cross-section and it can also be expressed as a percentage of the initial cross-section of tensile bars (see Table 4.4).

The same work can be conducted for the average cross-section loss between two service cracks (or between two shear stirrups). With the spacing between the cracks being around 200 mm and the spacing between the shear stirrups also 200 mm, a calculation is performed at intervals of 200 mm using the highest value of the crack openings recorded every 100 mm. The model by Khan *et al.* (equation [4.5]) is then used to perform the average cross-section (GC) loss calculation (see Table 4.5).

Figure 4.21 presents the plot of the possible amplitude of the corrosion pits along beam Bs04, for the two tensile bars, front and back, as well as the average of the local losses, LC, on the two tensile reinforcements, which thus corresponds to the local loss on a beam cross-section. The local loss, LC, for each bar may be correlated to the loss of ductility of the reinforcements and the average on the cross-section to the residual bearing capacity of the corroded element (see Chapter 6).

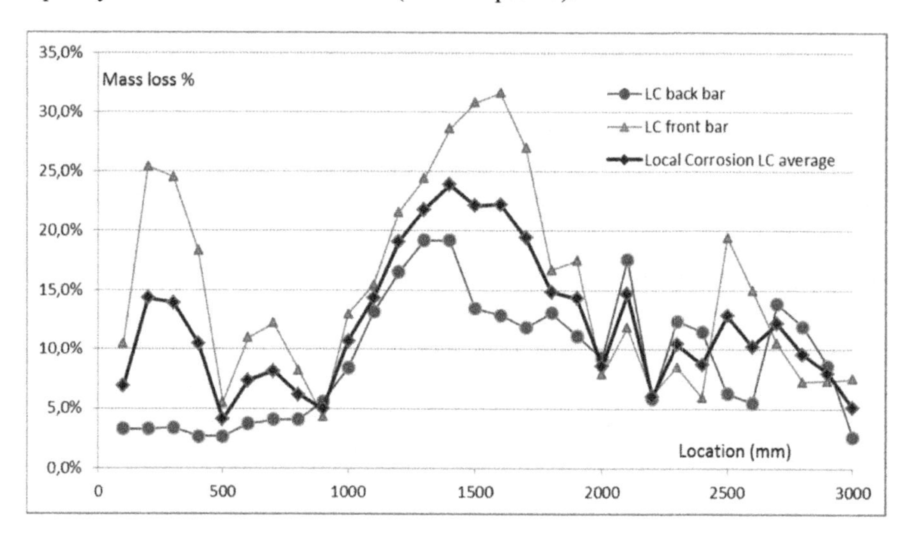

Figure 4.21. *Plot of the possible amplitude of the corrosion pits along beam Bs04, for the two tensile bars, front and back, as well as the average of the local losses, LC, on a beam cross-section. For a color version of the figure, see www.iste.co.uk/francois/corrosion.zip*

Figure 4.22 presents the plot of the average cross-section loss, GC, of reinforcements by 200-mm coupon (spacing between the shear stirrups and the service cracks, where applicable) due to corrosion along beam Bs04, for the two tensile bars, front and back, as well as their average value on a beam cross-section. The average loss, GC, may be used to assess the change in stiffness of the corroded elements and the growth of deflections in service (see Chapter 6).

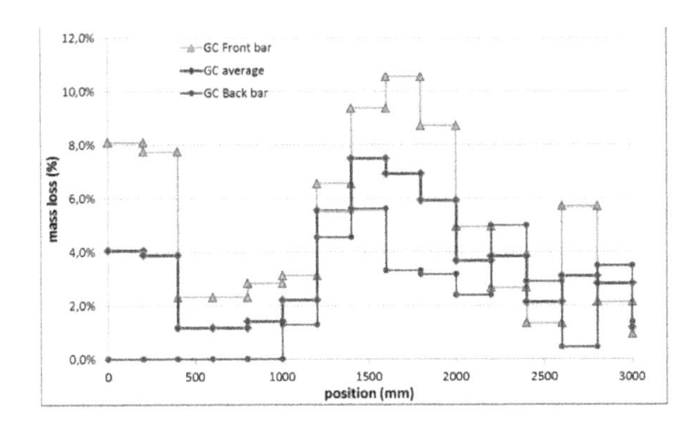

Figure 4.22. *Plot of the average cross-section loss, GC, by 200-mm coupon (spacing between the shear stirrups and the service cracks, where applicable) due to corrosion along beam Bs04, for the two tensile bars, front and back, as well as the average of the average losses, GC, on a beam cross-section. For a color version of the figure, see www.iste.co.uk/francois/corrosion.zip*

Figure 4.23 presents the comparison between the local, LC, and average, GC, cross-section losses for the tensile reinforcements of beam Bs04. As this figure clearly shows, the average loss, GC, is not deduced from the local losses, LC; they are two distinct models that evolve with respect to the corrosion crack openings, but which give two distinct and complementary pieces of information: the maximum possible local loss (LC) and the average generalized cross-section loss (GC) along a length of the order of magnitude of 100 to 200 mm.

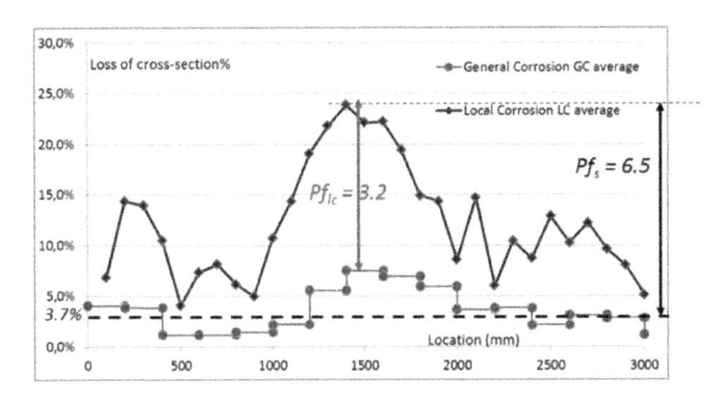

Figure 4.23. *Comparison between local cross-section loss, LC, and average cross-section loss, GC, by segment of 200 mm due to the corrosion of beam Bs04. It should be noted that the average loss is not directly correlated with local loss: these are two distinct models. For a color version of the figure, see www.iste.co.uk/francois/corrosion.zip*

Two of the possible definitions of pitting factor are also to be found in Figure 4.23:

– Firstly, factor pf_s, which corresponds to the ratio between maximum local loss (LC) and the average corrosion (GC) along the length of the beam; here it has a value of 6.5, which is consistent with the values obtained through experimentation (Figures 3.13 and 3.14).

– Secondly, factor pf_{lc}, which corresponds to the ratio between the local corrosion (LC) and generalized corrosion at the same location; in the example presented, it has a value of 3.2, which is consistent with the short duration of the corrosion process.

5

Effects of Reinforcement Corrosion on the Mechanical Behavior of Reinforced Concrete

5.1. Introduction

The purpose of this chapter is to present the effects of reinforcement corrosion on the mechanical behavior of reinforced concrete. The corrosion of reinforcements leads to a reduction in the resistant cross-section of the bars (which modifies the constitutive relation of steel in tension) and to a modification of the steel-concrete bond phenomenon owing to the de-confinement of the bars, linked to the creation of the corrosion cracks: these two phenomena alter the mechanical behavior of reinforced concrete in terms of both bearing capacity and deflections in service and deflections at failure.

We will therefore be studying successively in this chapter the influence of corrosion on the mechanical behavior of steel reinforcements, on steel-concrete bond and lastly on the behavior of the reinforced-concrete composite.

The study of corrosion effects on mechanical behavior is based on the results of *natural corrosion* tests on steel in concrete, nevertheless certain comparisons can be made with results of accelerated corrosion, which while not representative of the corrosion process, remains used by the majority.

5.2. Effect of corrosion on the mechanical behavior of reinforcement steel

The constitutive relation of steel for reinforced concrete is a local law, even if this relation is identical all along a non-corroded reinforcement. During the corrosion phenomenon, the occurrence of corrosion areas can be noted that are localized or not so localized along the reinforcements. The constitutive relation of steel will thus vary along the reinforcement according to the intensity of the local corrosion.

The constitutive relation of non-corroded steel is well known and its experimental characterization is produced by a straightforward tensile test, instrumented by strain gauges or displacement sensors if the extent of the yielding threshold is of interest. The measurement basis for the tensile extension is generally ten times' the diameter of the steel tested. The same measurement protocol is used for corroded steel (see Figure 5.1).

Figure 5.1. *Simple tensile test on a corroded steel bar [ZHU 14]. For a color version of the figure, see www.iste.co.uk/francois/corrosion.zip*

5.2.1. *Mechanical behavior of non-corroded steel*

A typical curve of the experimental behavior of the steel used for concrete reinforcements is presented in Figure 5.2. This steel, which is 12 mm in diameter and type Fe500, has been extracted from a non-corroded reinforced-concrete beam kept for 26 years in laboratory conditions, to serve as a control sample with respect to beams corroded during the same period.

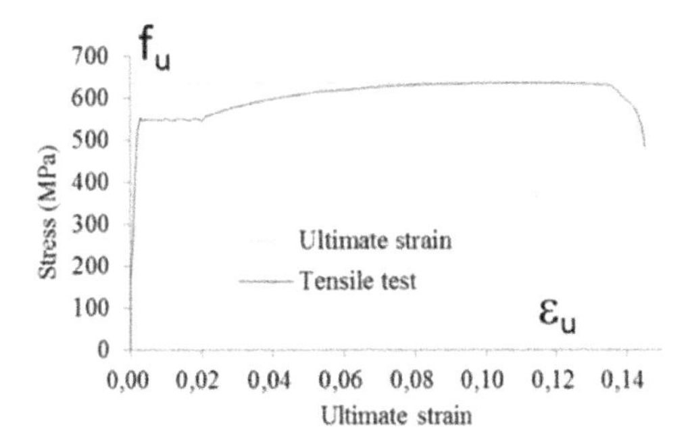

Figure 5.2. *Experimental constitutive relation under simple tension of a non-corroded steel reinforcement [ZHU 17a]. For a color version of the figure, see www.iste.co.uk/francois/corrosion.zip*

During a controlled tensile test in displacement, imposed on a non-corroded steel reinforcement, the following can be noted, successively: an elastic area, a plateau corresponding to the yielding stress, f_y, then a slight work hardening up to maximum stress at failure, f_u, with which is associated strain at maximum stress, ε_u. Beyond the maximum stress, the field of inhomogeneous plastic deformation is entered: this is the necking phenomenon, corresponding to the occurrence of a marked local reduction of the bar cross-section and determining the failure area. In order to maintain the imposed displacement set-point, the tensile force then decreases owing to the local reduction in cross-section, as well as the stress, which is calculated with respect to the nominal steel cross-section. It should be noted that the actual cross-section of the bar decreases over the course of the test (Figure 5.3), which leads to an increase in the actual stress throughout the tensile test.

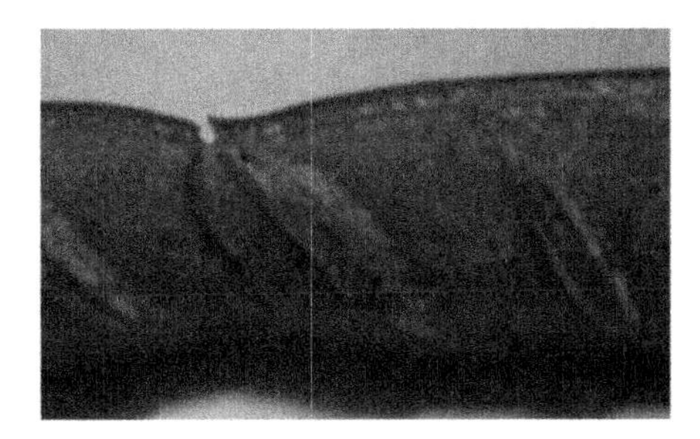

Figure 5.3. *Necking phenomenon prior to failure under simple tension of a non-corroded steel reinforcement [ZHU 13]. For a color version of the figure, see www.iste.co.uk/francois/corrosion.zip*

An important property of reinforced concrete is avoiding *brittle fractures*. There is no explicit recommendation verification for this, but an implicit verification where both a minimum percentage of steel reinforcement and sufficient ductility of the latter are known [EUR 92]. Table 5.1 recalls the minimum characteristics of steel reinforcements according to Eurocode 2. Three categories exist, with category C concerning seismic areas where a significant ductility is required, category B covers cases where redistribution of the stresses required after yielding is above 20%, and category A covers all other cases.

Properties of steel reinforcements (EC2 - Table C.1)			
Category	A	B	C
Yield strength, f_y (MPa)	400–600		
Minimum value of f_u/f_y	≥1.05	≥1.08	≥1.15 and ≤1.35
Strain at maximum stress, ε_u (%)	≥2.5	≥5.0	≥7.5

Table 5.1. *Properties of reinforcements, according to Eurocode 2: Appendix C – Table C.1*

For non-corroded reinforcements used in the long-term program conducted at INSA Toulouse [FRA 94a, FRA 94b, FRA 94c] (see the Appendix) and serving as experimental data for this chapter, the yield strength is 550 Mpa, the maximum stress of 640 MPa (corresponding to a f_u/f_y ratio of 1.16) and the corresponding elongation at the maximum stress of 10%. Thus, the reinforcements belong to category C, defined by Eurocode 2.

5.2.2. *Mechanical behavior of corroded steel*

Concerning the mechanical behavior of corroded steel, the corrosion characterization parameter is inevitably local: nevertheless, as seen in Chapter 3, this parameter is difficult to measure very accurately. The experimental results presented in this section correspond to steel reinforcements extracted from two naturally corroded beams, B2CL2 and B2CL3, tested by Zhu and François [ZHU 13, ZHU 14]. Figure 5.4 shows a cross-section comparison between a corroded steel reinforcement and a non-corroded steel reinforcement and the difficulty in characterizing the residual cross-section of corroded steel owing to the non-uniform corrosion pattern. Thus, the minimum residual diameter measurement leads to underestimation of the residual cross-section (or overestimation of the corrosion).

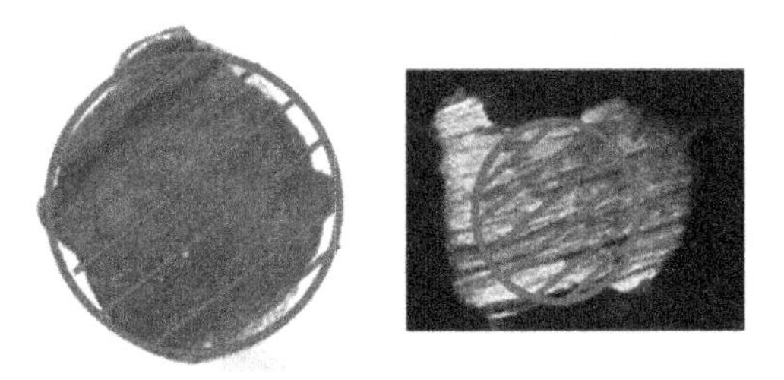

Figure 5.4. *Comparison between the cross-sections of a non-corroded steel reinforcement and those of a corroded steel reinforcement [ZHU 13]. For a color version of the figure, see www.iste.co.uk/francois/corrosion.zip*

Given this difficulty in calculating the residual cross-section after corrosion, Figure 5.5 shows a few typical corroded steel tensile test experimental curves in the force-displacement plane. It can be observed that the corrosion modifies the response of tensile steel with, in particular, the disappearance of the transition plateau between the yield behavior and the work-hardening phase. It reduces very extensively the lengthening at failure, without, in principle, any correlation with the decrease in tensile force at failure.

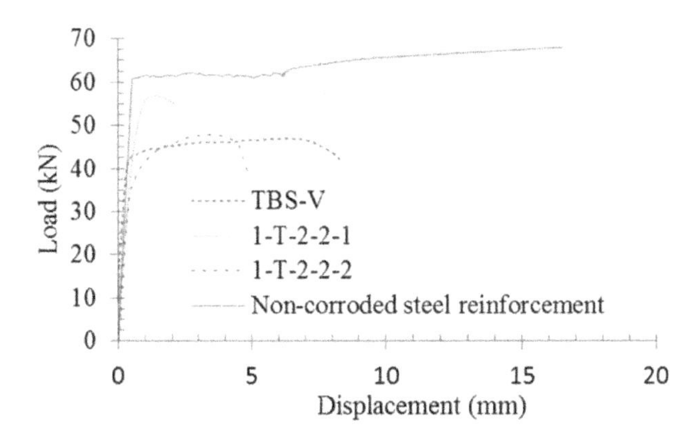

Figure 5.5. *Comparison between the typical behaviors of a corroded steel reinforcement and those of a non-corroded tensile steel reinforcement [ZHU 13]. For a color version of the figure, see www.iste.co.uk/francois/corrosion.zip*

Based on the residual local cross-section deduced from a gravimetric measurement as described in Chapter 3, it is possible to plot the response of corroded steel to a tensile test in the stress-strain plane.

5.2.2.1. *Effect of corrosion on yield strength*

The absence of a plateau in the stress-strain curve requires the use of the stress value calculated for a deformation of 0.2% as the yield strength. The strain, ε_u, for the maximum stress can also be deduced (see Figure 5.6).

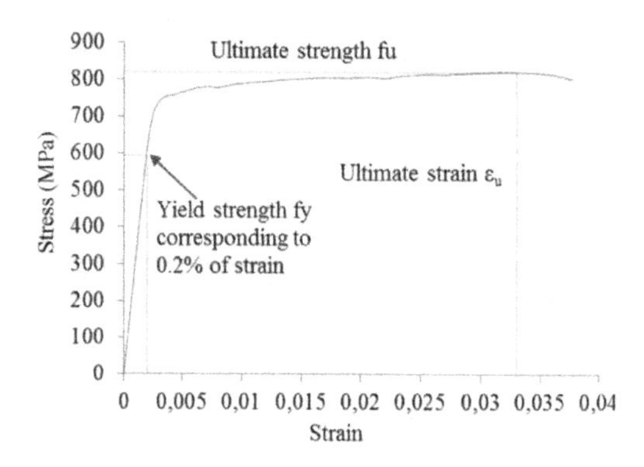

Figure 5.6. *Yield-strength calculation for a corroded steel reinforcement and strain corresponding to the maximum stress [ZHU 14]*

Calculating the yield strength, the maximum stress and the elongation at maximum stress enables the plotting of the evolution of these three amounts for different local corrosion-induced cross-section losses. Figure 5.7 shows that the yield strength is characterized by a certain variability due to inaccuracies on the residual cross-section measurement, but that these calculated values are close to those of non-corroded steel reinforcement, with an average of 550 MPa. It can be concluded from the above that *the steel yield strength is not affected by corrosion*, which is a logical result given that corrosion leads to a reduction in cross-section but not to a change in the properties of the steel.

5.2.2.2. *Effect of corrosion on maximum stress*

Again on Figure 5.7, it can be noted that the maximum stress reached during the tensile test is characterized by significant variability, but that the average value of around 780 MPa is substantially higher than that measured on non-corroded steel (640 MPa).

The steel reinforcement has not become stronger after corrosion, rather, there is a calculation bias linked to the change in ductility of the steel before and after corrosion. Indeed, in the absence of corrosion, the steel demonstrates a necking phenomenon characteristic of ductility, which leads to an increase in the actual failure stress with respect to that conventionally calculated using the nominal cross-section. In contrast, the corroded steel shows a brittle fracture and no necking. If we consider, like Zhu and François [ZHU 13], that the actual steel cross-section is overestimated by 20% when the necking phenomenon is not taken into account, then the actual ultimate stress of non-corroded steel reinforcement is obtained by increasing by 20% the apparent value of 640 MPa, i.e. 768 MPa, to be reconciled with the value of 780 MPa obtained as the ultimate stress of corroded steel reinforcements. In conclusion, *the ultimate steel stress is not modified by corrosion* but its value does not correspond to that of a conventional calculation based on the nominal cross-section of the bars.

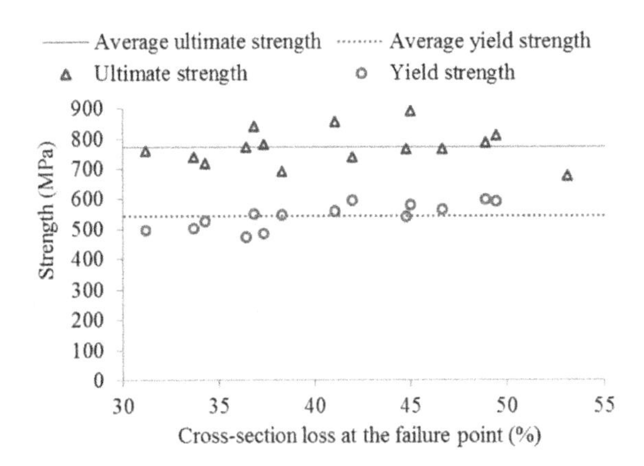

Figure 5.7. *Evolution of the yield strength and of the maximum stress for a corroded steel reinforcement with respect to the corrosion-induced cross-section loss [ZHU 14]. For a color version of the figure, see www.iste.co.uk/francois/corrosion.zip*

5.2.2.3. *Effect of corrosion on steel ductility (elongation reached at maximum stress)*

Figure 5.8 shows the evolution in elongation ratio at maximum stress measured on corroded steel reinforcements, with respect to that of non-corroded steel reinforcements. A very significant variability is observed that cannot simply be linked to the residual cross-section calculation, but above all a very significant decrease in the elongation at maximum stress, which can reach more than 99% of the elongation of non-corroded steel. The steel reinforcement that was initially category-C in the sense of Eurocode 2 with respect to ductility, is no longer even category-A in many cases after corrosion. This parameter is a significant indicator of *the steel-reinforcement ductility loss due to corrosion*.

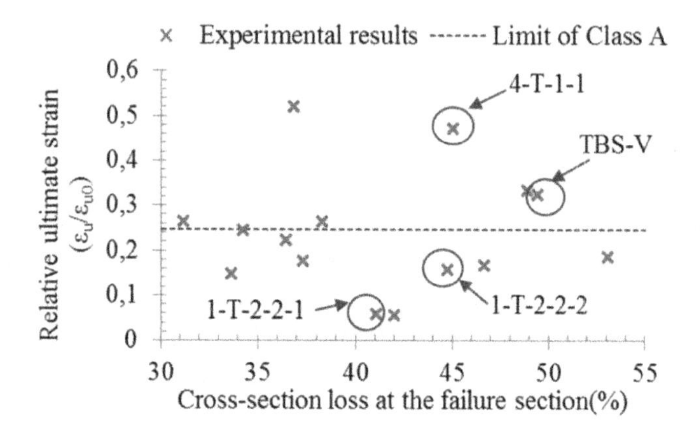

Figure 5.8. *Evolution of the relative elongation at maximum stress (ductility loss) with respect to the corrosion-induced cross-section loss [ZHU 17a, ZHU 17b]. For a color version of the figure, see www.iste.co.uk/francois/corrosion.zip*

The ductility loss illustrated in Figure 5.8 is linked to the cross-section loss generated by corrosion but also to the pattern (to the shape) of this cross-section loss. Thus, Figure 5.9 shows the failure of a corroded steel bar (1-T-2-2-1) with a non-uniform morphology of the corrosion pit on the reinforcement perimeter.

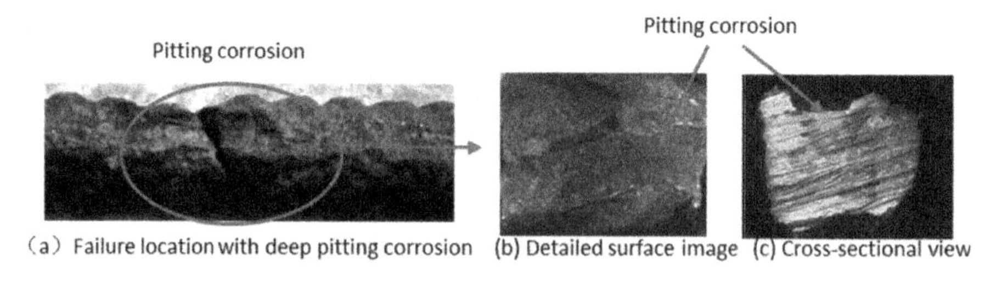

Figure 5.9. *View of the morphology of a corrosion pit, non-uniform around the perimeter and with a lengthways extension that is greater than at depth, as signaled in Chapter 1, according to Zhu* et al. *[ZHU 17a, ZHU 17b]. For a color version of the figure, see www.iste.co.uk/francois/corrosion.zip*

It can also be observed in Figure 5.9 that the necking phenomenon disappears in the presence of a corrosion pit. The mechanism that decreases the ductility of the steel is linked to the notch effect, which leads, by the effect of stress concentration at the pitting, to a partial yielding of the cross-section for low loads, then limiting the yield reservation necessary in order to obtain a significant ductility. The notch effect is linked to the cross-section loss shape. Thus, Zhu *et al.* [ZHU 17a, ZHU 17b] demonstrate the influence of the corrosion pit shape on ductility by simulating different forms of corrosion: uniform around the perimeter, UC, unilateral notch, P1, and symmetrical notch, P2 (see Figure 5.10). The pit morphology is quantified by a form factor, k_p, expressed empirically by Zhu *et al.* [ZHU 17a, ZHU 17b] and appearing to be correlated to the radius of gyration of the cross-section of the corroded bar.

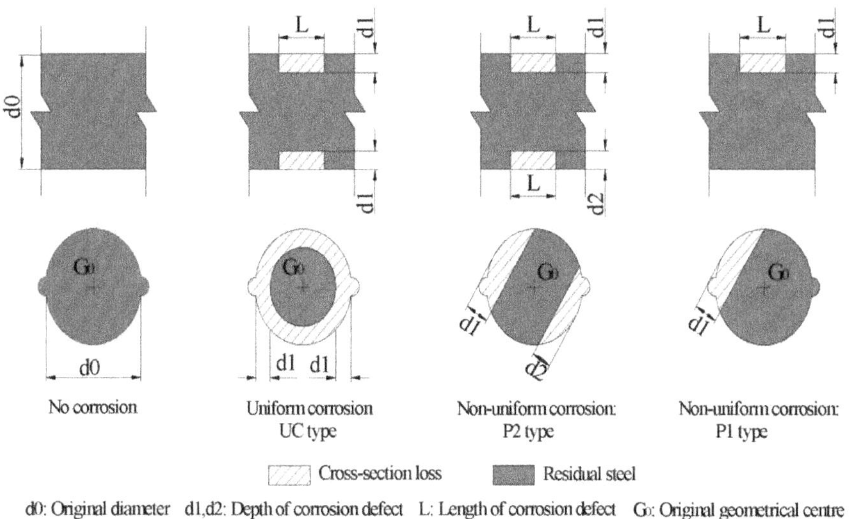

Figure 5.10. *View of the different notch shapes used to simulate corrosion pits to model the steel ductility loss with respect to the corrosion-induced cross-section loss [ZHU 17a, ZHU 17b]. For a color version of the figure, see www.iste.co.uk/francois/corrosion.zip*

Figure 5.11 presents the results of test performed on the idealized corrosion shapes, UC, P1 and P2. It completes them with the results from the literature [BAL 16, OU 16], including accelerated corrosion which, although not representative of natural corrosion, leads to more uniform cross-section losses, which is useful in order to confirm the importance of the local effects of natural corrosion.

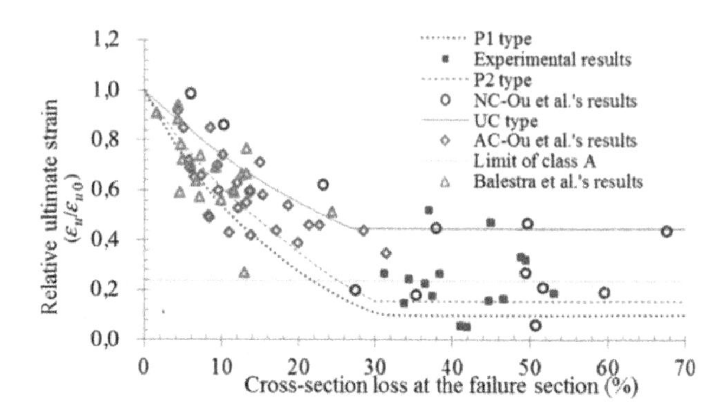

Figure 5.11. *Evolution of steel ductility loss with respect to the different notch types (UC, P1 and P2) in comparison with the results from the literature, according to Zhu* et al. *[ZHU 17a, ZHU 17b]. For a color version of the figure, see www.iste.co.uk/francois/corrosion.zip*

The steel ductility loss due to corrosion is a very significant characteristic for reassessing corroded structures. A model of this ductility loss with respect to the steel cross-section loss and the pitting form factor, k_p, is proposed by Zhu *et al.* [ZHU 17a, ZHU 17b], improving on a previous model by Castel *et al.* [CAS 00b].

$$\frac{\varepsilon_{u\_corr}}{\varepsilon_{u0}} = e^{-3k_pC} \quad \text{for } 0 < C \leq 0.3 \qquad [5.1]$$

$$\frac{\varepsilon_{u\_corr}}{\varepsilon_{u0}} = e^{-3k_p 0.3} \quad \text{for } C \geq 0.3 \qquad [5.2]$$

whereby:

– ε_{u_corr} = elongation at maximum stress of the corroded steel reinforcement.

– ε_{u0} = elongation at maximum stress of the non-corroded steel reinforcement.

– k_p = pitting form factor: k_p = 1 for a homogeneous corrosion on the perimeter (carbonation), $2 \leq k_p \leq 2.4$ for chloride-induced pits.

– C = steel cross-section loss $0 \leq C \leq 1$.

NOTE.– The mechanical behavior of corroded steel is modified by the presence of corrosion: disappearance of the plastic plateau leading on from the yield strength and very significant change in the failure mode, which becomes brittle with the disappearance of the necking phenomenon.

The yield strength, f_y, remains unchanged.

The maximum *actual* stress at failure is significantly higher than the maximum *nominal* stress at failure prior to corrosion.

This difference is due to the calculation mode: nominal cross-section for the non-corroded steel reinforcement and actual cross-section for the corroded steel reinforcement. The actual failure stress almost certainly remains unchanged. To this end, a failure calculation of a corroded element, based on the use of the maximum stress of the non-corroded steel reinforcement, will be safe, in underestimating the actual failure value.

5.3. Effect of corrosion on steel-concrete bond

Steel-concrete bond is an essential characteristic for the bending stiffness of reinforced-concrete elements and therefore mainly concerns the calculation of the displacements of deflected elements. This then presents the case of loads that are greater than the cracking load and therefore the presence of service cracks.

5.3.1. *Bond between non-corroded steel and concrete*

For a bent reinforced-concrete element, two phases in the formation of service cracks are differentiated: the formation phase and the stabilization phase [CEB 99]. At a crack, the steel strain is maximum as only steel resists the tensile forces. The bond phenomenon enables the gradual transmission of part of the force resisted by the steel to the concrete over a length known as the transfer length, L_t, [CEB 99, FRA 06b]. Beyond the transfer length, the steel and concrete experience a perfect bond. As a consequence, if the transfer length is less than half the distance between two primary bending cracks (general example), there is the occurrence of a secondary crack between the primary cracks (see Figure 5.12), reducing the spacing between cracks and leading to cracking stabilization. There can exist cases of the spacing between the primary cracks being less than twice the transfer length [PIY 04]. In this case, there is a single, directly stabilized cracking phase.

The transmission of part of the tensile force to the concrete between the cracks contributes to the bending stiffness of the reinforced-concrete elements and is known as the *Tension-Stiffening Effect*. Let us note that this force is entirely resisted by the steel at the cracks. Taking this phenomenon into account enables the deflections of elements subjected to bending to be predicted. Thus, the standards offer general approaches based on the introduction of an equivalent tie surface in the cracked area of the tensile beams [CEB 99].

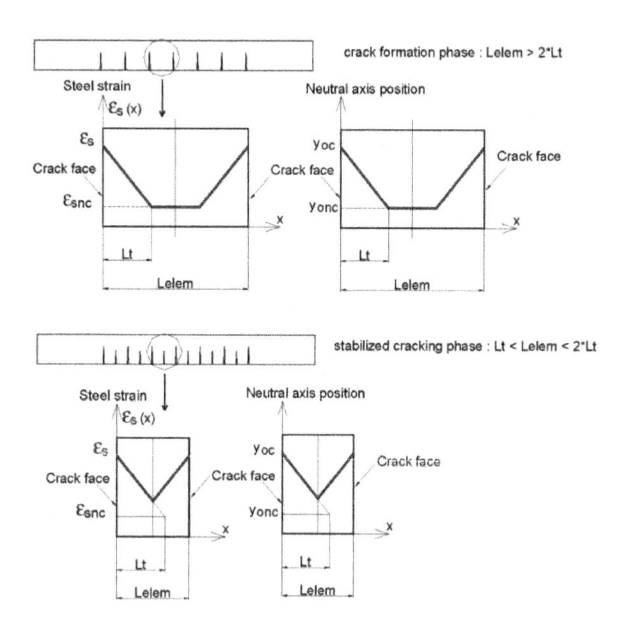

Figure 5.12. *Phases of crack formation, then stabilization, and evolution of strains of the steel and neutral axis between the bending cracks, according to Vu* et al. *[VU 10]. For a color version of the figure, see www.iste.co.uk/francois/corrosion.zip*

5.3.2. *Bond between corroded steel and concrete*

The corrosion process in a reinforced concrete structure exposed to the atmosphere leads to the creation of expansive corrosion products and, consequently, cracks along the longitudinal reinforcements (Chapter 4). The creation of corrosion cracks leads to de-confinement of the tensile reinforcements and therefore to a reduction in the bond between steel and concrete. This bond reduction leads to an increase in the transfer length, which becomes L_{tcor}, and a decrease in the stiffening effect of the tensile concrete (see Figure 5.13) [FRA 06b].

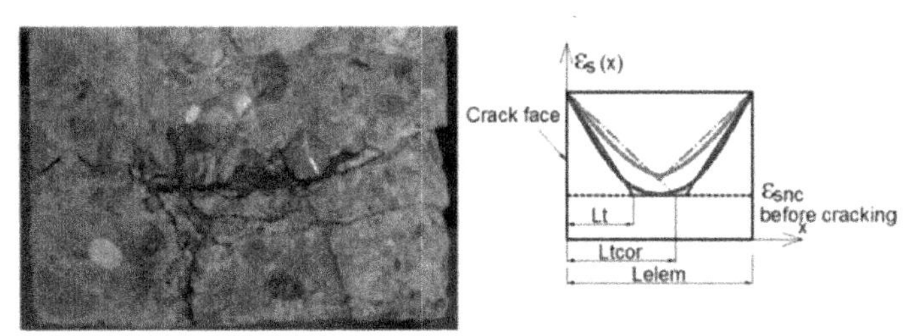

Figure 5.13. *De-confinement of tensile reinforcements owing to the creation of corrosion cracks and increase in the transfer length between reinforcements and concrete [FRA 06b]. For a color version of the figure, see www.iste.co.uk/francois/corrosion.zip*

With the increase in corrosion and the progressive de-confinement of the bars, it can even be possible to obtain a cancellation of the steel-concrete bond and thus an infinite transfer length (see Figure 5.14). To avoid handling a parameter that can vary to infinite, it is preferable to introduce a bond damage variable, D_c, varying between 0 (no corrosion) and 1 (total bond loss) (equation [5.3]):

$$L_{tcor} = \frac{L_t}{1 - D_c} \quad [5.3]$$

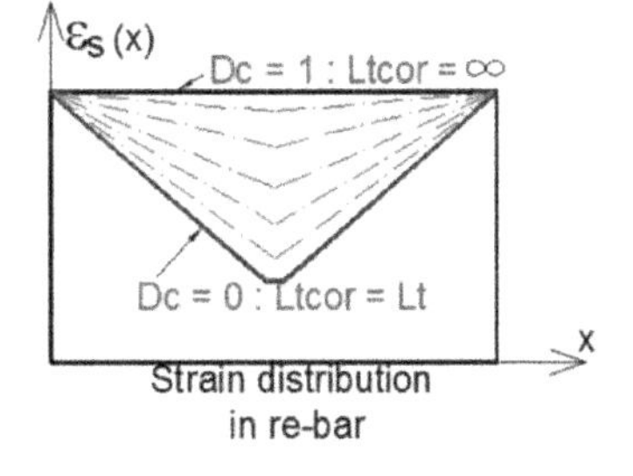

Figure 5.14. *View of the variation in transfer length with respect to the corrosion intensity and the corresponding damage variable [FRA 06b]. For a color version of the figure, see www.iste.co.uk/francois/corrosion.zip*

The deterioration in bond due to corrosion is linked to the creation of corrosion cracks, which themselves are linked to the cross-section loss of the reinforcements. The amount of corrosion products formed is thus the motor of de-confinement. The damage variable is therefore expressed with respect to the steel cross-section loss due to corrosion for a given reinforcement bar. François *et al.* [FRA 06b] propose to model the damage variable by taking into account the reinforcement cross-section loss, only if it is above the corrosion crack creation threshold, ΔA_{s0} (Chapter 4) (equation [5.4]):

$$D_c = 1 - \left(\frac{A_s - \Delta A_s}{A_s - \Delta A_{s0}} \right)^5 \qquad [5.4]$$

if $\Delta A_s < \Delta A_{s0}$, otherwise $D_c = 0$

This damage variable depends on the initial diameter of the bars through parameter A_s (reinforcement cross-section), and does not take into account the compensating effect of the shear stirrups to reduce the de-confinement due to the creation of corrosion cracks. An improvement to this model was proposed by Castel *et al.* [CAS 16], which breaks away from the influence of the initial diameter and introduces the taking into consideration of the shear stirrups (equation [5.5]).

$$D_c = \min\left(\frac{2}{pf_g} st \log(1 + 0.25(\Delta A_s - \Delta A_{s0}); \; 1 \right) \qquad [5.5]$$

if $\Delta A_s < \Delta A_{s0}$, otherwise $D_c = 0$

whereby pf_g is the geometric pitting factor, and st the shear-stirrup confinement parameter: $st = 1$ in the absence of shear stirrups, otherwise $st < 1$ is calculated by equation [5.6],

$$st = 1 - 1.8 \frac{A_{st}}{A_s} \left(1 - 0.64 \frac{A_{st}}{A_s} \right) \qquad [5.6]$$

whereby A_{st} represents the transverse reinforcement surface along the anchorage length of the longitudinal bars.

A comparison of the two models for taking into account corrosion-induced bond loss via a damage variable is presented in Figure 5.15.

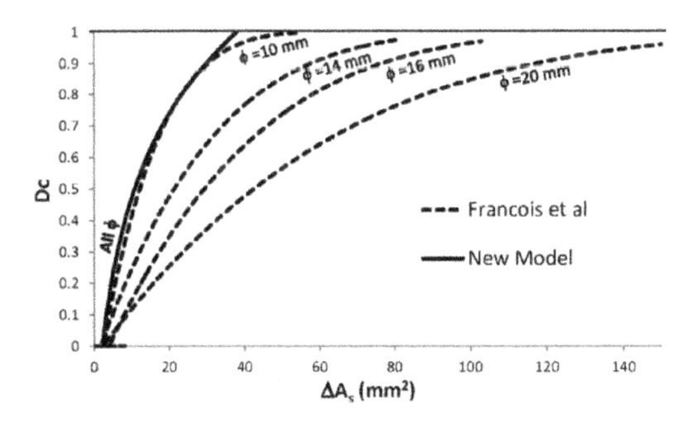

Figure 5.15. *Comparison of the steel-concrete bond damage variable models: models by François* et al. *[FRA 06b] and Castel* et al. *[CAS 16]*

The damage variable, D_c, which enables quantification of the effect of corrosion on the bond, may also be directly expressed as a relative loss in bond stress (equation [5.7]), which enables its use in other types of reinforced-concrete mechanical-behavior modeling, using finite interface elements.

$$D_c = \frac{\tau_{u0} - \tau_{uc}}{\tau_{u0}}, \qquad [5.7]$$

whereby τ_{u0} and τ_{uc} are the bond stresses before and after corrosion, respectively.

A validation of this approach is presented by Castel *et al.* [CAS 16] based on a comparison with results obtained under natural corrosion by Tahershami *et al.* [TAH 14] on elements extracted from Stallbacka Bridge in Sweden after 30 years of service. The longitudinal reinforcements measured 16 mm and a shear stirrup of 10 mm was present on the anchorage length. The model by Sajedi and Huang [SAJ 15] was also compared for its ability to take account of the shear stirrups, even if it does not distinguish non-uniform, natural corrosion from uniform, accelerated corrosion. Despite a substantial variability in the experimental results, the model by Castel *et al.* with a pitting parameter of $pf_g = 4$ demonstrates correct predictions, which are conservative and therefore safe (see Figure 5.16).

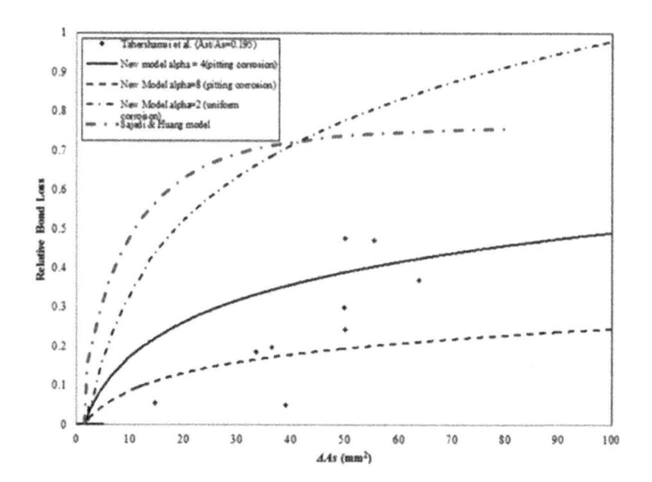

Figure 5.16. *Comparison between the experimental results of Tahershami* et al. *[TAH 14] and the steel-concrete bond damage variable models by Castel* et al. *[CAS 16] and Sajedi and Huang [SAJ 15]. For a color version of the figure, see www.iste.co.uk/francois/corrosion.zip*

Taking D_c into account in the calculation of the transfer length calculation, L_{tcor}, enables the deflections of reinforced-concrete elements to be predicted (Figure 5.17): this calculation is detailed in Chapter 6.

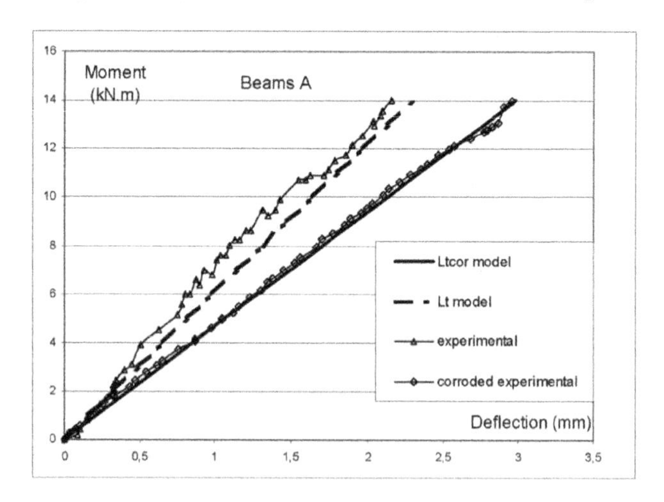

Figure 5.17. *Comparison between the experimental and calculated deflections of a beam corroded naturally for 17 years (A1CL1) and a control beam, taking into account the bond loss due to corrosion via the damage variable, Dc, according to François* et al. *[FRA 06b]*

5.4. Effect of corrosion on the bending mechanical behavior of reinforced-concrete beams

Corrosion reduces reinforcement cross-section, modifies the constitutive equation of tensile steel in terms of ductility and decreases steel-concrete bond: the mechanical behavior at structure scale is therefore also affected. We will first observe the effect of corrosion on behavior at ULS: yielding load, ultimate load and deflection at failure of reinforced-concrete beams. This influence at ULS is mainly linked to maximum local corrosion (LC), which is predominant at the start of the corrosion process. We will then examine the influence of corrosion on behavior at SLS: change in stiffness and deflections in service. This influence at SLS is linked to both local corrosion (LC), and the generalization of corrosion (GC) along the corrosion cracks, which leads to cross-section losses distributed along the reinforcements.

5.4.1. *Effect of corrosion on behavior at ULS: yielding load, bearing capacity and deflection at failure*

Figure 5.18 shows comparisons between the typical bending response of non-corroded beams and of corroded beams of the same age but with different time periods (14, 23, 26 and 28 years). The reinforced-concrete beams, aged in saline environment under sustained load, presented as an example of the behavior of corroded reinforced concrete, are summarized in Table 5.2. The corrosion indicated in Table 5.2 is the maximum local corrosion (LC) in the maximum stress area.

It should be noted that the tested beams were kept under service load during aging (including the controls) and therefore no long demonstrate the behavior before cracking in Figure 5.18. The non-corroded beams have a typical behavior: yielding of the tensile steel reinforcements and failure of the compressed concrete, which assures satisfactory ductility and sufficient alert signals prior to failure. In contrast, the corroded beams present a reduced yield threshold and a sudden and brittle fracture consecutive to the failure of the corroded tensile reinforcements.

Author	Date	Ref.	Conservation	Max. cross-section loss (LC)	Yielding threshold loss	Ultimate capacity loss	Ductility loss
Castel *et al.* [CAS 00a]	2000	B1CL 1	Cl^- 14 years	20%	19%	16.7%	70.7%
Zhang *et al.*	2009	B2CL 1	Cl^- 23 years	36%	38.3%	27.5%	52.7%
[ZHA 09a, ZHA 09b]		B2T	notches 23 years	30%	30%	26.5%	80%
Zhu and François	2013	B2CL 2	Cl^- 26 years	34%	34.4%	27.3%	50.2%
[ZHU 15a]		B2CL 3	Cl^- 28 years	43%	43.1%	39.9%	73.2%

Table 5.2. *Summary of the studies conducted and results obtained on the mechanical behavior of reinforced-concrete beams cast in 1984 in Toulouse [YU 15b]*

Figure 5.19 shows, by way of illustration, the failure by crushing of the compressed concrete of control beam B2T, and that by failure of the tensile steel reinforcements (without compression failure of the concrete despite the presence of corrosion cracks in the compressed area), of beam B2CL3.

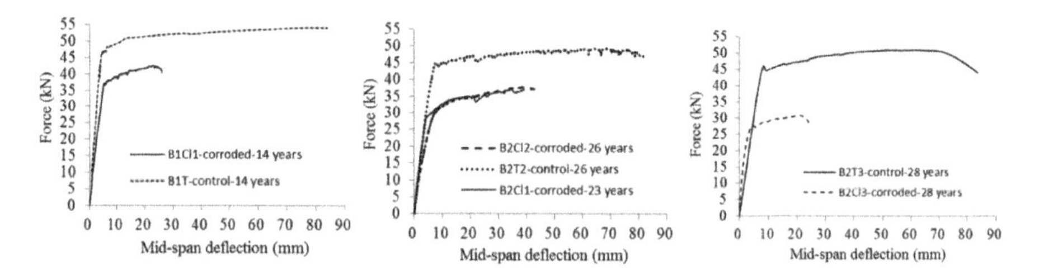

Figure 5.18. *Comparison between experimental bending behaviors of corroded and non-corroded beams for different corrosion periods, according to Zhu and François [ZHU 15a]*

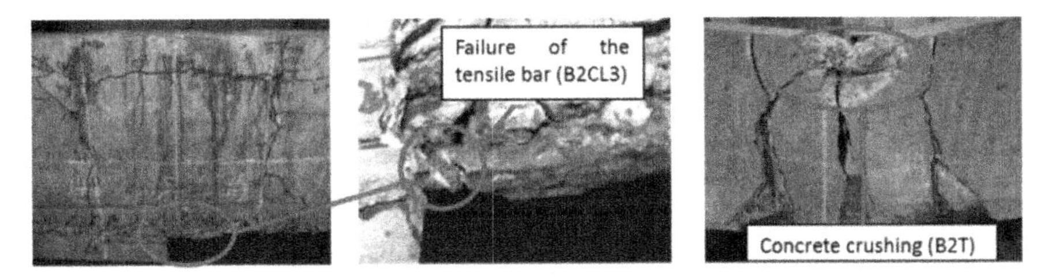

Figure 5.19. *Comparison of reinforced-concrete beam failure modes before and after corrosion, according to Zhu and François [ZHU 15a]. For a color version of the figure, see www.iste.co.uk/francois/corrosion.zip*

5.4.1.1. *Effect of corrosion on yielding load*

The load that leads to tensile-reinforcement yielding is, for a conventional calculation of reinforced concrete, directly proportional to the cross-section of tensile reinforcements. Having seen that the steel yield strength was not modified by corrosion, if it is possible to measure the residual steel cross-section of the corroded reinforcements in the failure area, it is then found that the yielding load loss is proportional to the cross-section loss (Table 5.2), which is displayed in Figure 5.20.

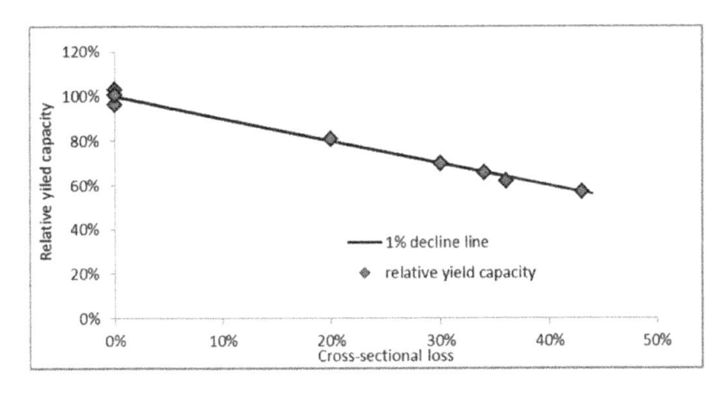

Figure 5.20. *Correlation between decrease in yielding load and loss of steel cross-section due to corrosion, according to Yu* et al. *[YU 15b]. For a color version of the figure, see www.iste.co.uk/francois/corrosion.zip*

5.4.1.2. *Effect of corrosion on ultimate load or bearing capacity*

Where bearing capacity or ultimate load are concerned, the prediction is more complex than that of yielding load, as depending on the concrete damage under compression, the steel work-hardening phenomenon and the steel ductility reduction, which modifies the failure mode. This complexity also comes from the fact that the ultimate stress used for the calculation does not have exactly the same meaning for sound steel as for corroded steel: nominal stress and actual stress, respectively. Figure 5.21 shows the decrease in bearing capacity with respect to the steel cross-section loss for the same beams as those of Figure 5.20.

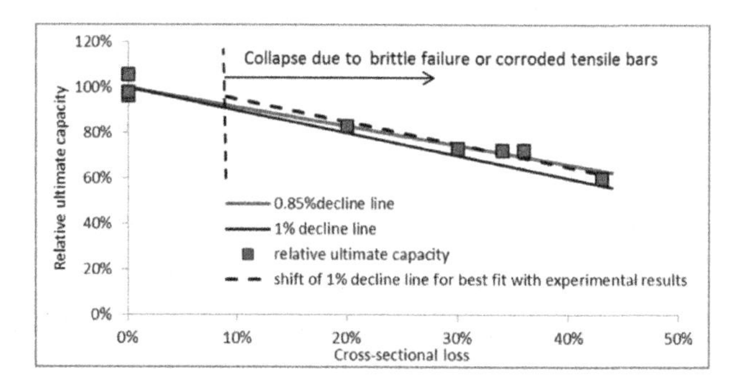

Figure 5.21. *Correlation between decrease in bearing capacity and loss of local steel cross-section due to corrosion, according to Yu* et al. *[YU 15b]. For a color version of the figure, see www.iste.co.uk/francois/corrosion.zip*

In Figure 5.21 it can be noted that the loss of bearing capacity is close to the proportionality to local steel cross-section loss at the failure area. With the failure mode evolving from the crushing of compressed concrete to the failure of tensile steel reinforcements, it is logical to find a proportionality as soon as the cross-section loss due to corrosion leads to failure of the tensile reinforcements, as seen in Figure 5.21, beyond 10% cross-section loss (value relative to the elements tested).

The decrease in bearing capacity appears to be correlated with the local steel cross-section loss due to corrosion, which is an important result for the prediction of the bearing capacity of corroded structures during a recalculation diagnosis. Despite the non-representative nature of corrosion said to be accelerated under electrical field, if we are in a position to assess the local cross-section loss at the failure area, we find results comparable with natural corrosion. This comparison is presented in Figure 5.22. Beware that this is only true for bearing capacity assessment taking the local cross-section loss at the most stressed cross-section. It is not comparable, however, in terms of failure mode and failure elongation.

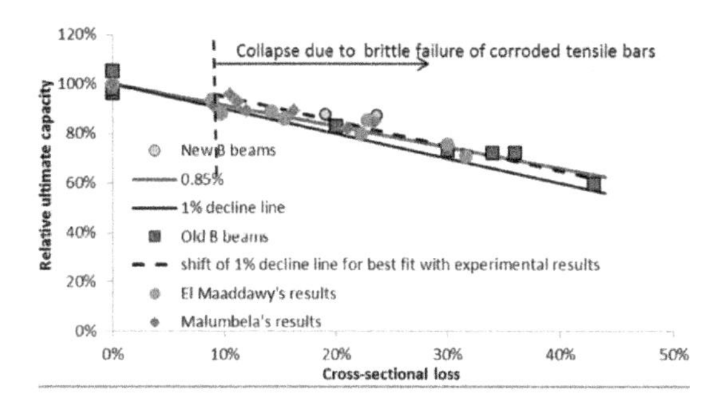

Figure 5.22. *Correlation between decrease in bearing capacity and loss of local steel cross-section due to corrosion for different corrosion types: natural or accelerated, according to Yu* et al. *[YU 15b]. For a color version of the figure, see www.iste.co.uk/francois/corrosion.zip*

5.4.1.3. *Effect of corrosion on maximum deflection at failure characteristic of ductility, and ductility index of the corroded elements*

We saw in section 5.2.2.3 that corroded steel ductility was very significantly reduced according to the steel cross-section and the form factor, k_p, of the pitting. This reduction in ductility modifies the bending failure mode, which becomes brittle, with failure of the tensile steel reinforcements at the corrosion pits. Thus we obtain a correlation between the brittle fracture of corroded steel reinforcements and the brittle fracture of corroded reinforced-concrete beams, characterized by a reduced deflection at failure that potentially presents a danger to goods and people. Figure 5.23 shows the change in bending behavior due to corrosion for four corroded beams: A2CL3 aged 26 years and A2CL1, A2CL2 and A1CL2 aged 27 years, reported by Dang and François [DAN 14]: it is clear in Figure 5.23 that the change in ductility due to corrosion is very much identified by a very significant decrease in ultimate deflection.

For non-corroded reinforced-concrete elements, a ductility index exists that is defined as the ratio between the deflection reached at ultimate load with respect to the deflection reached at yielding load [ELC 03]. The difficulty when using this index in the case of corroded beams is that the corrosion modifies both the deflection at yielding and the deflection at maximum load, and that these two variations are antagonistic. They can therefore give rise to contradictory or inconsistent results (see Figure 5.24(a)) bearing no relation to the brittle fractures observed experimentally. To alleviate this difficulty, Dang and François [DAN 14] and Zhu and

François [ZHU 14] defined a corrosion ductility index (*CDI*) for reinforced-concrete beams equaling the ratio of the maximum deflection at failure of a corroded beam compared to that of a non-corroded beam (see Figure 5.24(b)). The corrosion ductility index appropriately describes the decrease in ductility of the reinforced-concrete elements owing to corrosion.

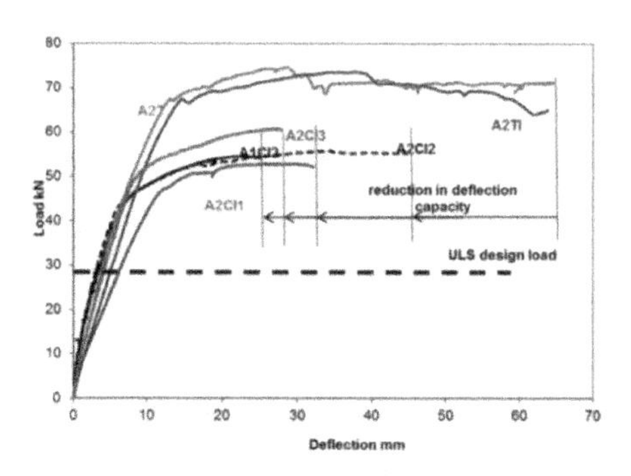

Figure 5.23. *Effect of corrosion on the mechanical behavior of reinforced-concrete beams and in particular on ultimate deflection, according to Dang and François [DAN 14]. For a color version of the figure, see www.iste.co.uk/francois/corrosion.zip*

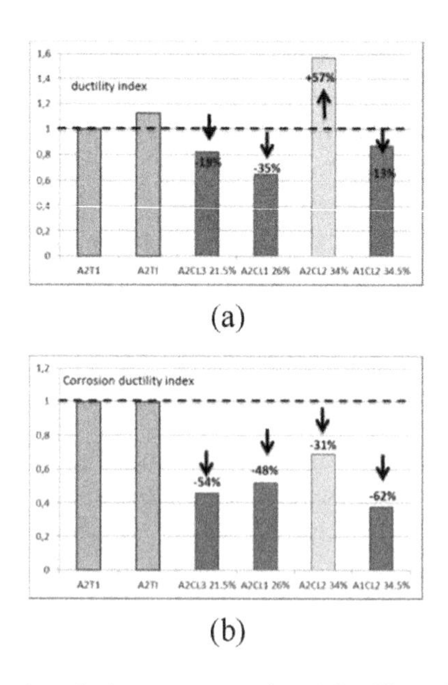

Figure 5.24. *Comparison between conventional ductility index a) and corrosion ductility index b), introduced by Dang and François [DAN 14]. For a color version of the figure, see www.iste.co.uk/francois/corrosion.zip*

NOTE.– Ductility reduction due to corrosion cannot be viewed using the conventional ductility index, which refers to yield behavior. It is necessary to focus directly on the reduction in ultimate deflection via the corrosion ductility index, which takes account of the, more fragile, change in behavior of reinforcements at failure.

5.4.2. *Effect of corrosion on behavior at SLS: modification of deflections in service, redistribution of forces by steel yielding*

5.4.2.1. *Effect of corrosion on deflections in service*

The localized initial corrosion (LC) of the reinforcements leads to the creation of corrosion cracks, which modify the steel-concrete bond and therefore the deflection in service of the reinforced-concrete elements. Pursuing the corrosion process leads to a generalization of the corrosion (GC) along the corrosion cracks, therefore leading to their extension along the reinforcements and to an increase in their openings (Chapter 4). The influence on deflections in service is then due to the overlapping of the bond loss and the average cross-section loss along the reinforcements (GC). This process is illustrated in Figure 6.25, which demonstrates the evolution of the deflection at mid-span for a corroded beam, B2CL1, at different time periods: 14, 19 and 23 years. Observing the evolution of corrosion cracks (Figure 5.26) between the three time periods, it can be noted that the extension of the corrosion crack length has a substantial impact on deflection, with a significant increase between 14 years and 19 years, whereas the increase in very significant crack openings between 19 and 23 years barely modifies the deflection.

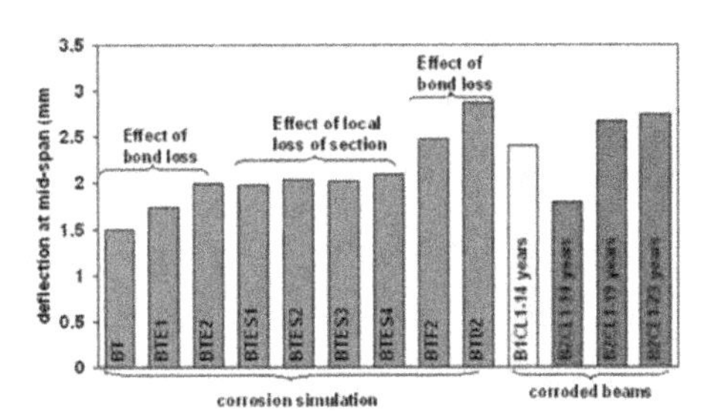

Figure 5.25. *Deflections at mid-span measured on corroded beams B1CL1 and B2CL1 at different stages in the corrosion process and comparison with different stages in simulations of bond losses and local cross-section losses, according to Zhang* et al. *[ZHA 09b]. For a color version of the figure, see www.iste.co.uk/francois/corrosion.zip*

The main cause of the modification of deflections in service is therefore linked to the modification of the steel-concrete bond induced by corrosion, whereas the influence of local corrosion (LC) is negligible. This point is also verified by Zhang *et al.* [ZHA 09b] by simulating a bond loss between steel and concrete by partially removing the concrete around the reinforcement and by simulating localized corrosion by creating notches in the reinforcements (Figure 5.27). In Figure 5.25, BTE1 and BTE2 correspond to the stripping over the distance between two bending cracks for a tensile reinforcement, then for the two tensile reinforcements of control beam BT. BTES1 to BTES4 correspond to the creation of corrosion notches (25% local loss, LC), and BTF2 to the stripping on a second sector between two bending cracks and BTD2 to the stripping on a third sector between two bending cracks. It is clear in Figure 5.25 that each bond loss leads to a significant increase in deflection whereas the notches have only very little effect.

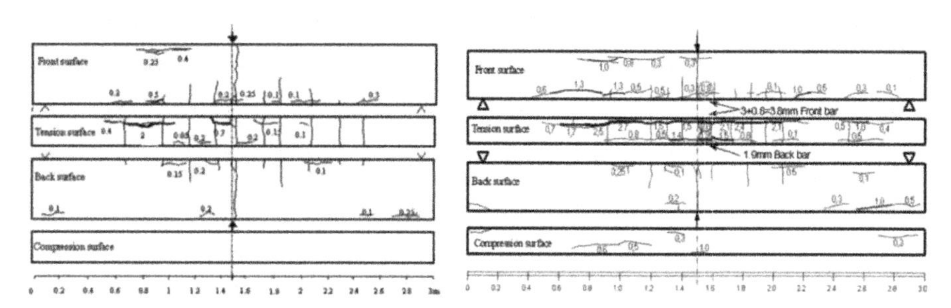

B2CL1: 14 years of corrosion B2CL1: 19 years of corrosion

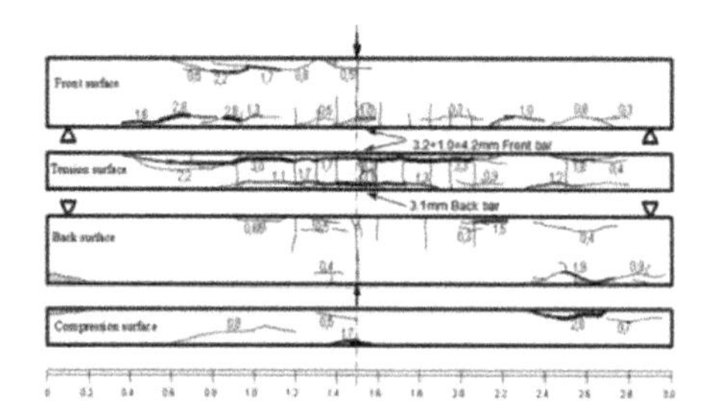

B2CL1: 23 years of corrosion

Figure 5.26. *Evolution of corrosion-induced cracks along beam B2CL1 at time periods of 14, 19 and 23 years, according to Zhang* et al. *[ZHA 09b]*

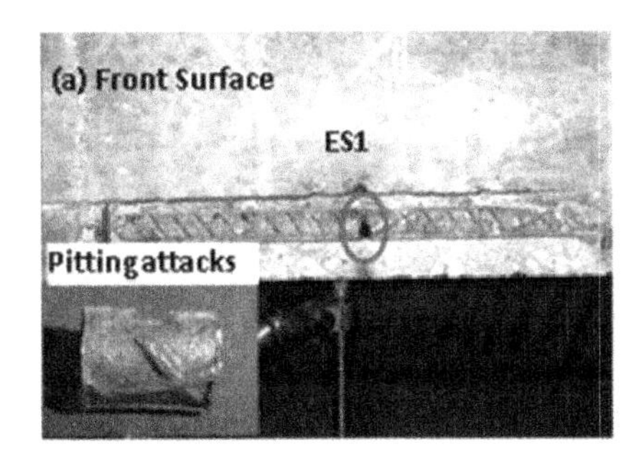

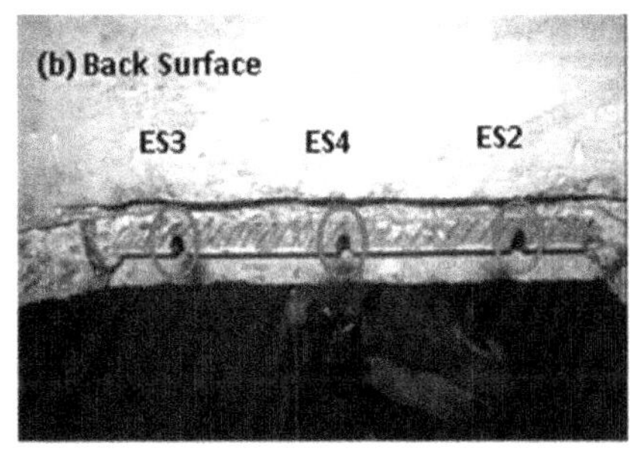

Figure 5.27. *Simulation of bond loss by removing the cover concrete and simulation of localized corrosion by creating notches on a control beam, BT, according to Zhang* et al. *[ZHA 09b]. For a color version of the figure, see www.iste.co.uk/francois/corrosion.zip*

5.4.2.2. *Effect of corrosion on redistribution of forces by steel yielding*

Statistically-indeterminate, reinforced-concrete structures can reach a higher ultimate load than that resulting from a conventional calculation at ULS thanks to the force redistribution mechanism permitted by the formation of one or more plastic hinges (Figure 5.28).

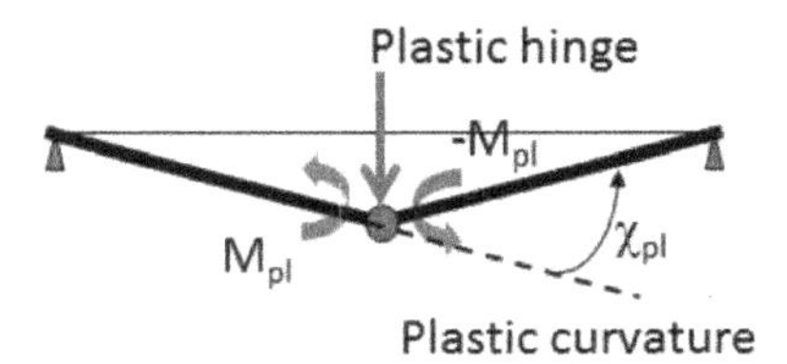

Figure 5.28. *Schematic diagram of a plastic hinge for a simple beam. For a color version of the figure, see www.iste.co.uk/francois/corrosion.zip*

The formation of one or more plastic hinges, in this case statistically-indeterminate structures, authorizes large displacements: this aspect is particularly important in the case of seismic loads. Localized corrosion, which is reflected by a sharp decrease in reinforcement ductility, significantly reduces the possibility of plastic hinge formation. *It is to be noted that the accelerated tests found most often in the literature (corrosion said to be under electrical field) do not enable the scale of this embrittlement phenomenon to be observed* as they lead to more homogeneous corrosion on the perimeter and the length of the reinforcement bars.

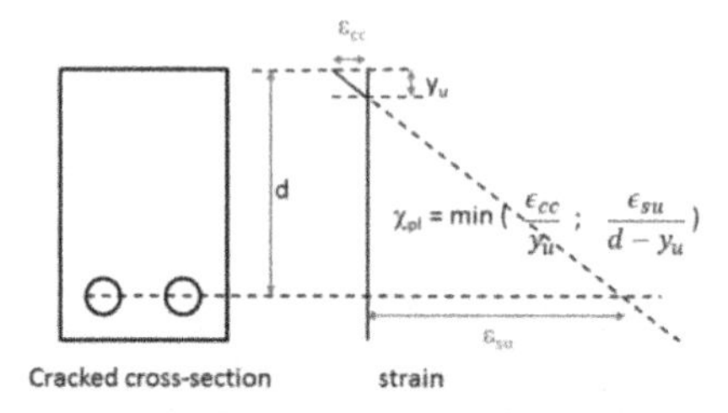

Figure 5.29. *Determination of the bending limit of a plastic hinge for the case of a failure of compressed concrete or tensile concrete. For a color version of the figure, see www.iste.co.uk/francois/corrosion.zip*

Figure 5.29 recalls the determination of the ultimate bending limit of a plastic hinge for the case of a compressed-concrete failure corresponding to the case of non-corroded reinforced concrete (equation [5.8]),

$$\chi_{pl} = \frac{\varepsilon_{cc}}{y_u} \qquad [5.8]$$

and in the case of the failure of tensile steel reinforcements that can correspond to corroded reinforced concrete (equation [5.9])

$$\chi_{pl} = \frac{\varepsilon_{su}}{d - y_u} \qquad [5.9]$$

NOTE.– The corrosion decreases the ultimate strain, ε_{su}, and therefore the admissible plastic bending. In the presence of corrosion, it is no longer possible to consider a redistribution of forces by reinforcement yielding.

The principle of calculating deflection at failure in the case of a plastic hinge will be presented in Chapter 6.

5.5. Effect of corrosion on mechanical behavior with regard to the shear force of reinforced-concrete beams

The behavior of reinforced-concrete structural elements with regard to the shear force is very complex and it is difficult to separate, in practice, the shear behavior from the bending behavior, as they are always coupled.

Studying the long beams presented in the above section (5.4) enables us to consider that the influence of the shear force is negligible. On the contrary, the study of short beams increases the shear-force influence but does not eliminate the bending influence.

This section will therefore solely present a few test results on naturally corroded short beams. These tests show that the presence of corrosion on the tensile reinforcements, by decreasing the bending strength capacity, can modify the failure mode of short beams, which evolves from an initial brittle fracture due to the shear force, to a bending failure leading on from the tensile-steel failure. These tests were conducted by Zhu *et al.* [ZHU 15b] on four short beams in 2013 (see Table 5.3).

Authors	date	Ref.	Length, mm	span, mm	State
	2013	B2T3-1	1,020	16.7%	control
	2013	B2T3-2	1,020	26.5%	control
Zhu and François	2013	B2CL3 -1	1,200	27.3%	corroded
[ZHU 15b]	2013	B2CL3-2	1,040	39.9%	corroded

Table 5.3. *Summary of the characteristics of short beams, studied by Zhu* et al. *[ZHU 15b]*

The control beams have failed in a diagonal shear failure mode with a brittle collapse following the formation of inclined cracks (Figure 5.30).

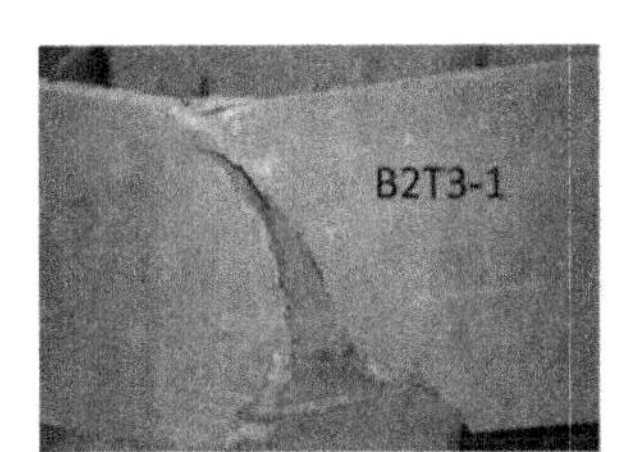

Figure 5.30. *View of the failure mode at shear force of control beams B2T3-1 and B2T3-2 [ZHU 15b]. For a color version of the figure, see www.iste.co.uk/francois/corrosion.zip*

The corroded beams have failed in bending mode with a brittle fracture due to failure of the corroded tensile steel reinforcements (Figure 5.31).

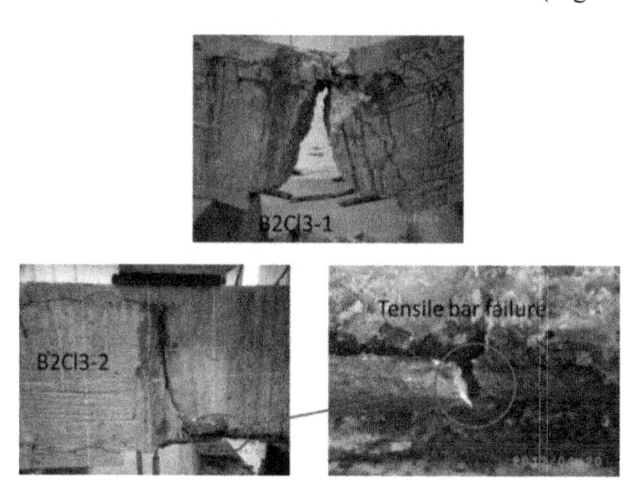

Figure 5.31. *View of the failure mode at* shear force *of corroded beams B2Cl3-1 and B2Cl3-2 [ZHU 15b]. For a color version of the figure, see www.iste.co.uk/francois/corrosion.zip*

The force-deflection curves of the control and corroded beams confirm the respective failure modes by shear force and by bending (Figure 5.32). The corroded beams demonstrate a yield threshold typical of bending failures and for forces below those of the control beams. The control beams show a brittle fracture at shear force for significantly higher load values than the corroded beams.

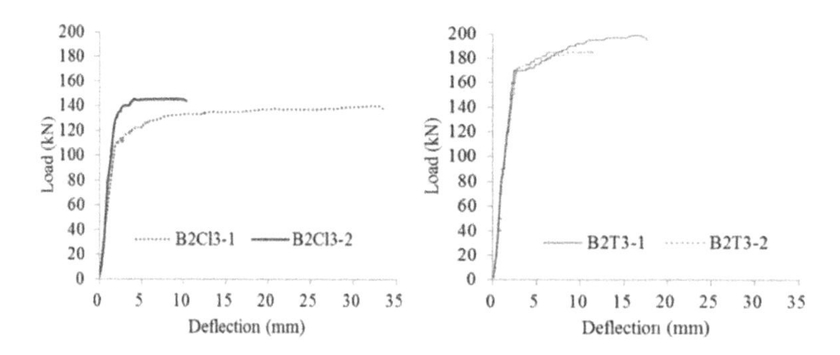

Figure 5.32. *Force-deflection curves of corroded beams B2Cl3-1 and B2Cl3-2 and control beams B2T3-1 and B2T3-2 [ZHU 15b]. For a color version of the figure, see www.iste.co.uk/francois/corrosion.zip*

It should be noted that neither the control beams nor the corroded beams showed slipping of the tensile reinforcements on supports, despite the reinforcements being deprived of anchorage hooks and in spite of the presence of corrosion-induced longitudinal cracks at the end of the corroded beams. The confinement brought by the support reactions and the shear stirrups enable a bond and sufficient anchorage of the tensile bars to be maintained.

6

Prediction of Bearing Capacity and Behavior in Service of Corroded Structures

6.1. Introduction

The aim of this chapter is to present an assessment method using the calculation of corroded structures. Using the diagnosis of corrosion of a structure based on the analysis of the cracks created by corrosion (Chapter 4), we propose to analyze the behavior of the structure and to determine its residual bearing capacity as well as its residual ductility, and therefore its capacity to redistribute force through the formation of a plastic hinge.

The first part of this chapter will be dedicated to the bearing capacity, deflection at failure and ductility of a reinforced-concrete beam, taking into account a bending load. The case of shear force that was simply mentioned in terms of a change in the mode of corrosion-induced failure, in the previous chapter, will not be studied in terms of prediction.

The second part of this chapter will be dedicated to operation in service, with the calculation of the new displacement field of a corroded beam subject to a simple bending load.

This chapter will be restricted to the calculation of a bent deflected element (beam). The consideration of a combined bending and compression (beam-column) will be easily deduced from that of the bent element by superposing a behavior under compression that is not affected by the corrosion of steel reinforcements.

The practical application will be conducted on beam Bs04, stressed under simple bending (Figure 6.1), for which the corrosion-induced crack mapping is presented in Chapter 4.

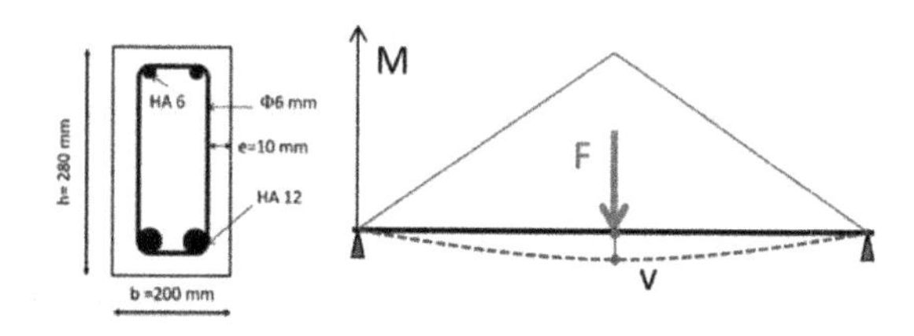

Figure 6.1. *Steel-reinforcement layout and stress mode of beam Bs04. For a color version of the figure, see www.iste.co.uk/francois/corrosion.zip*

6.2. Framework for assessing the actual (residual) behavior of a corroded structure

One of the difficulties in the reassessment (or the actual, residual assessment) of a corroded structure's mechanical behavior is that no recommendations framework exists. It is nevertheless planned for, be it in the structural Eurocode standards ("assessment and renovation of existing structures") or in the "model-code" of the FIB, to propose such recommendations in the more or less near future.

The approach taken in this chapter will not be to adapt to recommendations calculations that existed at the time the structure to be recalculated was designed, but rather to perform as accurate a calculation as possible using present-day knowledge: it will thus be down to the engineer who conducts this recalculation to propose an estimate of the safety margin according to current practices.

6.3. Phenomenological study of the behavior of a reinforced-concrete beam under bending, before and after corrosion

The typical behavior of a reinforced-concrete beam is presented in Figure 6.2. The envelope curve of the response is characterized by three points: A, which corresponds to the tensile-concrete cracking load, B, which corresponds to the yielding of the tensile steel reinforcements, and C, which corresponds to the ultimate load. Situated between points A and B are the phases of cracking, formation and stabilization, and the bending stiffness of the beam can only be grasped by an (at least partial) unloading. In the event of loading up to point D or E (Figure 6.2), the corresponding bending stiffness values are written as k_1 and k_2.

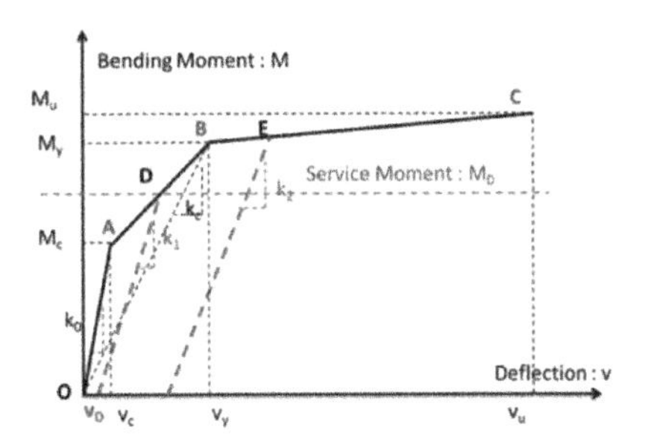

Figure 6.2. *Behavior of a non-corroded reinforced-concrete beam under bending. For a color version of the figure, see www.iste.co.uk/francois/corrosion.zip*

When a reinforced-concrete beam forms part of an existing structure, it can be assumed that the service loading has exceeded the cracking load, and the behavior until failure of such a beam before corrosion would then be described by the curve shown in Figure 6.3, with a linear behavior until historical loading (point D) then beyond, the behavior of the envelope curve identical to that shown in Figure 6.2. After corrosion, the behavior in service of the beam is modified: the deflection in service is enhanced by the corrosion effect (point D_{corr} in Figure 6.3) and the points representing the yielding load and ultimate load are also modified (points B_{corr} and C_{corr} in Figure 6.3). It should be noted that in Figure 6.3 the reloading is conducted as of the residual deflection, v_D, which would be obtained by fully unloading the beam. The deflections actually measured on the corroded beam are therefore relative to the residual value, v_D.

In the following sections, we will assess the position of the different representative points of Figures 6.2 and 6.3 enabling reassessment of the bearing capacity and ductility. We will also present a method for calculating bending stiffness before and after corrosion.

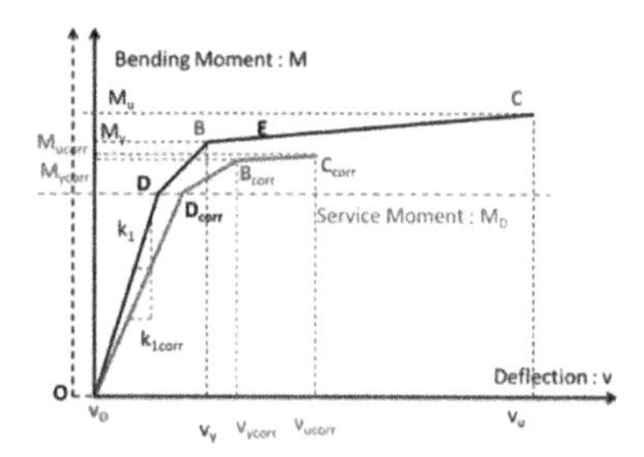

Figure 6.3. *Comparison between bending behaviors of a non-corroded reinforced concrete beam and a corroded beam. For a color version of the figure, see www.iste.co.uk/francois/corrosion.zip*

6.4. Assessment of the actual (residual) bearing capacity of a corroded structure

The geometrical data of the beam studied are as follows: b represents its width, h its height, e the concrete cover, d the effective height of its cross-section, c the distance from the tensile reinforcements to the beam surface, L the length, L_c the cracked length when the yielding moment is reached and A_s the transverse cross-section of the tensile steel reinforcements. Regarding the material properties, E_c is used for the Young's modulus of concrete, E_s for that of steel and n for the ratio between E_s and E_c.

It should be noted that the distance, c, can differ from the concrete cover. Indeed, the concrete cover corresponds to the minimum distance between the reinforcements and the concrete surface and is often relative to the shear stirrups. In the case of a longitudinal reinforcement, c then corresponds to the increased cover of the stirrup diameter.

6.4.1. *Behavior before corrosion*

To determine the envelope curve of the response under simple bending of beam Bs04, it is necessary to determine the position of points A, B and C in terms of moment and deflection.

As regards the behavior of the materials, the steel will be assumed to be perfect elasto-plastic and the constitutive relation of concrete will be taken to be of the parabola-rectangular type (Figure 6.4).

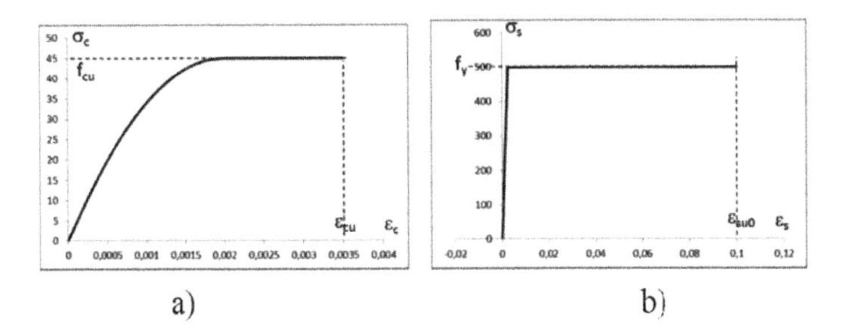

Figure 6.4. *Constitutive relations of concrete and steel*

6.4.1.1. *Determination of the cracking moment and the corresponding deflection*

The cracking moment, M_c, corresponding to point A, is calculated in terms of non-cracked cross-section (Figure 6.5). The position of the neutral axis, y_{0nc}, is determined by writing that the normal force is zero, then the moment M_c as the value resulting from the elementary moments of the stresses around the neutral axis for a value of the stress in the concrete that reaches its tensile strength, f_{ct}. Figure 6.5 presents the diagrams of strains and stresses in beam Bs04 before cracking. ε_{cc} is used for the compressed concrete's maximum strain, σ_{cc} for the corresponding stress, ε_{ct} for the maximum strain of the tensile concrete, σ_{ct} for the corresponding stress, ε_{cnc} for the strain of the tensile reinforcements and σ_{snc} for the corresponding stress. The bending curvature before cracking is written as χ_{nc}.

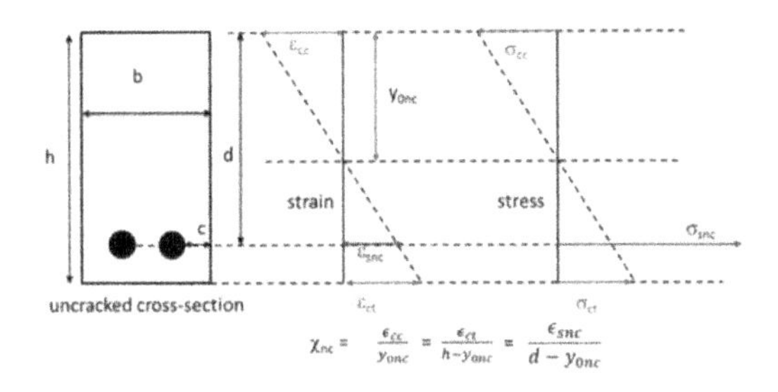

Figure 6.5. *Diagram of the stresses and strains in a beam cross-section prior to cracking. For a color version of the figure, see www.iste.co.uk/francois/corrosion.zip*

Equation [6.1] recalls the value of the non-cracked curvature with respect to the strains of the compressed concrete, the tensile concrete or that of the tensile reinforcements.

$$\chi_{nc} = \frac{\varepsilon_{cc}}{y_{0nc}} = \frac{\varepsilon_{ct}}{h - y_{0nc}} = \frac{\varepsilon_{snc}}{d - y_{0nc}} \quad [6.1]$$

Equation [6.2] gives the height of the neutral axis as a non-cracked cross-section resulting from writing the normal force as equal to zero.

$$y_{0nc} = \frac{\frac{bh^2}{2} + nA_s d}{bh + nA_s} \quad [6.2]$$

The bending moment is then conventionally written by the moment-curvature relation (equation [6.3]).

$$M = E_c I_{nc} \chi_{nc} \quad [6.3]$$

I_{nc} is the inertia of the non-cracked beam cross-section and is written according to equation [6.4].

$$I_{nc} = nA_s (d - y_{0nc})^2 + \frac{bh^3}{12} + bh(\frac{h}{2} - y_{0nc})^2 \quad [6.4]$$

The cracking moment, M_c, is obtained when the strain of the tensile concrete reaches its tensile limit value and the corresponding curvature is written as χ_{c0} (equations [6.5] and [6.6]).

$$M_c = E_c I_{nc} \chi_{c0} \quad [6.5]$$

$$\chi_{c0} = \frac{\frac{f_{ct}}{E_c}}{h - y_{0nc}} \quad [6.6]$$

The corresponding deflection, v_c, results from a conventional calculation for beam structures (equation [7.7]).

$$v_c = \frac{M_c L^2}{12 E_c I_{nc}} \quad [6.7]$$

6.4.1.2. *Determination of the yielding moment and the corresponding deflection*

The yielding moment, M_y, corresponding to point B is calculated in terms of cracked cross-section (Figure 6.6). The position of the neutral axis, y_{0c}, is determined by writing that the normal force is zero, then the moment M_y as the value resulting from the elementary moments of the stresses around the neutral axis for a tensile stress value in the tensile steel reinforcement that reaches its yield point, f_y. Figure 6.6 presents the diagrams of strains and stresses in a cracked cross-section of beam Bs04.

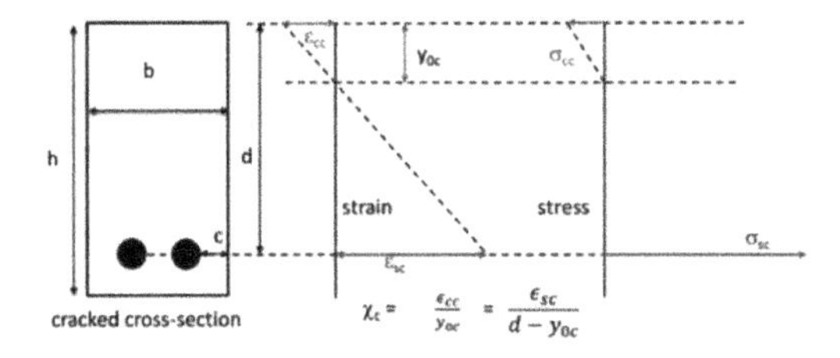

Figure 6.6. *Diagram of the stresses and strains in a cracked cross-section of beam below the yielding load. For a color version of the figure, see www.iste.co.uk/francois/corrosion.zip*

The bending curvature as a cracked cross-section is written as χ_c.

Equation [6.8] recalls the value of the cracked curvature with respect to the strains of the compressed concrete or the strain of the tensile reinforcements.

$$\chi_c = \frac{\varepsilon_{cc}}{y_{0c}} = \frac{\varepsilon_{sc}}{d - y_{0c}} \qquad [6.8]$$

Equation [6.9] gives the height of the neutral axis as a non-cracked cross-section resulting from writing the normal force as equal to zero.

$$y_{0c} = \frac{\left(-nA_s + \sqrt{n^2 A_s^{\,2} + 2nbA_s}\right)}{b} \qquad [6.9]$$

The bending moment is then conventionally written by the moment-curvature relation (equation [6.10]).

$$M = E_c I_c \chi_c \qquad [6.10]$$

I_c is the inertia of a cracked beam cross-section and is written according to equation [6.11].

$$I_c = nA_s(d - y_{0c})^2 + \frac{b y_{0c}^{\ 3}}{3} \qquad [6.11]$$

The yielding moment, M_y, is obtained when the strain of the tensile steel reaches its yield point, f_y, and the corresponding curvature is written as χ_y (equations [6.12] and [6.13]).

$$M_y = E_c I_c \chi_y \qquad [6.12]$$

$$\chi_y = \frac{\frac{f_y}{E_s}}{d - y_{0c}} \qquad [6.13]$$

The corresponding deflection, v_y, can be determined by conducting a conventional calculation for beam structures, considering that the bending stiffness, k_c, corresponding to the start of the steel-reinforcement yielding (slope of line 0B in Figure 6.2), reflects the bending response of the beam, calculated by allocating the cracked-cross-section inertia value all along the cracked length, L_c [VU 10].

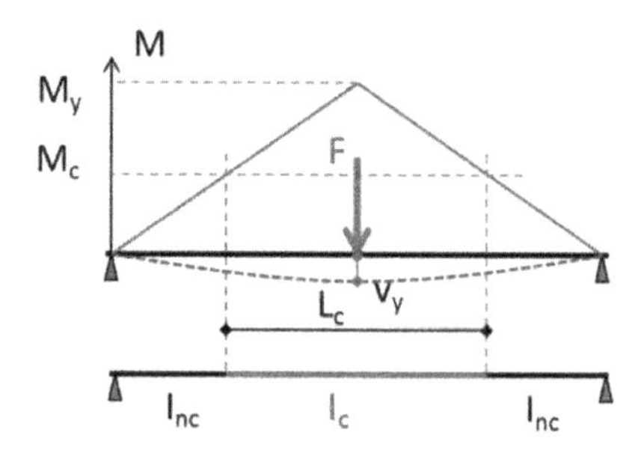

Figure 6.7. *Bending moment diagram of the beam corresponding to the yielding load of the median cross-section being reached: the length of the cracked area is written as L_C on the diagram. The whole cracked length is allocated cracked inertia for the calculation of the corresponding deflection when the yielding moment is reached. For a color version of the figure, see www.iste.co.uk/francois/corrosion.zip*

The whole cracked length, L_c, is allocated cracked inertia for the calculation of the corresponding deflection when the yielding moment is reached. The calculation of the deflection corresponding to the yielding of the steel reinforcements is therefore to be performed with a variable inertia value (Figure 6.7); nevertheless, a sufficient approximation can be obtained by considering the cracked inertia all along the beam. The deflection, v_y, is then obtained with equation [7.14].

$$v_y = \frac{M_y L^2}{12 E_c I_c} \qquad [6.14]$$

6.4.1.3. *Determination of the ultimate moment and the corresponding deflection*

The ultimate moment, M_u, corresponding to point C, is calculated as a cracked cross-section but taking account of the fact that the concrete is highly stressed, with a response that is approximated by a rectangular diagram over 80% of the compressed height, y_u (Figure 6.8) and a value of the stress in the concrete that is the same as its ultimate value, f_{cu}. The position of the neutral axis, y_u, is determined by writing that the normal force is zero, then the moment, M_u, as the value resulting from the elementary moments of the stresses around the neutral axis.

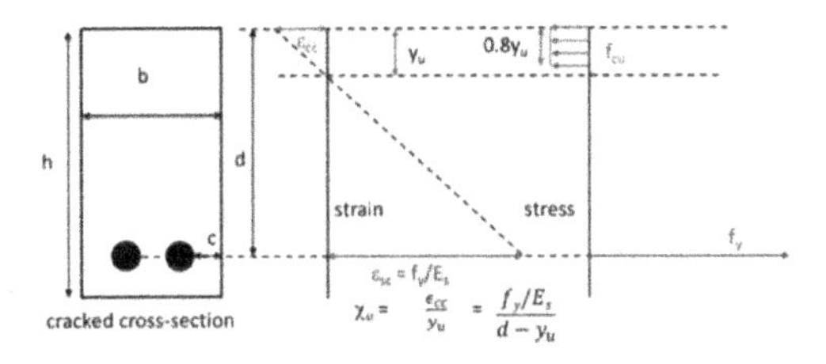

Figure 6.8. *Diagram of the stresses and strains in a cracked cross-section of beam beyond the steel yielding threshold. For a color version of the figure, see www.iste.co.uk/francois/corrosion.zip*

The bending curvature as a cracked cross-section is written as χ_u.

Equation [6.15] recalls the value of the cracked curvature with respect to the strains of the compressed concrete or the strain of the tensile reinforcements.

$$\chi_u = \frac{\varepsilon_{cc}}{y_u} = \frac{\frac{f_y}{E_s}}{d - y_u} \quad [6.15]$$

Equation [6.16] gives the height of the neutral axis as a non-cracked cross-section resulting from writing the normal force as equal to zero.

$$y_u = \frac{A_s f_y}{0{,}8 b f_{cu}} \quad [6.16]$$

The bending moment is then written as resulting from the compression forces in the concrete and tensile forces in the steel (equation [6.17]).

$$M_u = 0{,}48 y_u b f_{cu}^{\ 2} + A_s f_y (d - y_u) \quad [6.17]$$

The ultimate deflection, v_u, which is very important in assessing behavior ductility, can be determined based on the rotation capacity of the plastic hinge, which will be written as χ_{pl} formed after the yielding moment, by assessing whether it is limited by the maximum strain allowed under compression of the concrete ε_{cu} or by the maximum allowed extension of the tensile steel reinforcements ε_{su0} (equation [6.18]).

$$\chi_{pl} = \min\left(\frac{\varepsilon_{cu}}{y_u}; \frac{\varepsilon_{su0}}{d - y_u}\right) \quad [6.18]$$

The ultimate deflection, v_u, is calculated by approximating that the entire beam curvature beyond the formation of the plastic hinge is concentrated at the plastic hinge. The increase in deflection between the formation of the plastic hinge and the failure, v_u-v_y, is therefore proportional to the yield curvature (see Figure 6.9). The yield curvature is expressed with respect to the rotation, ϕ, of the two half-beam coupons by equation [6.19].

$$\chi_{pl} = 2\phi \quad [6.19]$$

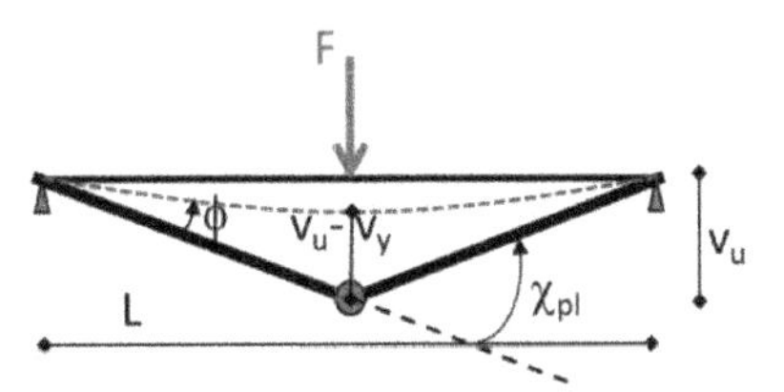

Figure 6.9. *Schematic diagram of the calculation of increase in beam deflection after the occurrence of a plastic hinge. For a color version of the figure, see www.iste.co.uk/francois/corrosion.zip*

It is then written that the increase in deflection is proportional to ϕ, according to equation [6.20].

$$\phi \approx \tan \phi = \frac{v_u - v_y}{\frac{L}{2}} \tag{6.20}$$

The beam's ultimate deflection at failure is written according to equation [6.21].

$$v_u = v_y + \frac{L}{4} \chi_{pl} \tag{6.21}$$

6.4.2. *Behavior after corrosion*

To determine the envelope curve of the response under simple bending of corroded beam Bs04, it is necessary to determine the position of points A_{corr}, B_{corr} and C_{corr} in terms of moment and deflection.

Regarding the material behavior, the concrete constitutive relation will remain unchanged (Figure 6.4(a)). In contrast, the steel constitutive relation is modified by the corrosion with a more fragile behavior, characterized by a reduction in the maximum relative extension at failure (Figure 6.10), using the model by Zhu *et al.* [ZHU 17a] (Figure 6.10(b)) presented in Chapter 5.

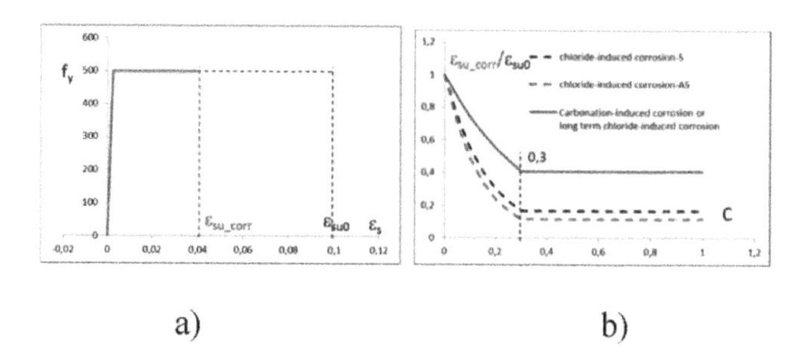

Figure 6.10. *Constitutive relation of corroded steel in comparison with non-corroded steel and influence of the reduction in cross-section due to corrosion, C (expressed from 0 to 1) on the decrease in the ratio of relative elongation at failure, for the case of chloride-induced (LC) or carbonation-induced corrosion (similar to the phase of chloride-induced corrosion generalization (GC)). For a color version of the figure, see www.iste.co.uk/francois/corrosion.zip*

Figure 6.10(b) shows the amplitude of the difference in ductility reduction between chloride-induced corrosion at the start of the propagation phase (essentially localized corrosion, LC) and that initiated by carbonation (via the pitting form coefficient, $_{p}u$, used in the model by Zhu *et al.* [ZHU 17a]). The generalization of the corrosion process due to the presence of cracks induced by corrosion (GC) may then be likened to corrosion due to carbonation.

To recalculate the envelope curve of the simple bending response of the corroded beam, the maximum local cross-section loss ΔA_s, taken from the model of localized corrosion (LC), is used. By subtracting ΔA_s from the initial cross-section, A_s, the residual steel cross-section, A_{s_corr} is obtained (equation [6.22]).

There would seem to be no point recalculating the cracking moment and the corresponding deflection of beam Bs04 (corresponding to point A_{corr}) after the corrosion process has begun, as given that the structures are corroded in service, they have already exceeded their cracking moment within the most stressed cross-sections. If, however, it should prove useful to perform this calculation, the steel cross-section, A_s, would simply need to be replaced in the equations of section 6.4.1.1 by the steel cross-section value, A_{s_corr}.

$$A_{s\_corr} = A_s - \Delta A_s \quad [6.22]$$

6.4.2.1. *Determination of the yielding moment and the corresponding deflection (corroded beam)*

The yielding moment, M_{y_corr}, corresponding to point B_{corr} is calculated in terms of cracked cross-section (Figure 6.6), as before corrosion. The position of the neutral axis, y_{0c_corr}, is determined by writing that the normal force is zero, then the moment, M_{y_corr}, as the value resulting from the elementary moments of the stresses around the neutral axis for a value of the tensile stress in the tensile steel that reaches its yield strength, f_y.

Equation [6.23] gives the height of the neutral axis as a non-cracked cross-section resulting from writing the normal force as equal to zero.

$$y_{0c\_corr} = \frac{\left(-nA_{s\_corr} + \sqrt{n^2 A_{s\_corr}^2 + 2nbA_{s\_corr}}\right)}{b} \qquad [6.23]$$

I_{c_corr} is the inertia of the cracked and corroded beam cross-section, written according to equation [6.24].

$$I_{c\_corr} = nA_{s\_corr}(d - y_{0c\_corr})^2 + \frac{by_{0c\_corr}^3}{3} \qquad [6.24]$$

The yielding moment, M_{y_corr}, is obtained when the stress of the tensile steel reaches its yield point, f_y, and the corresponding curvature is written as $\chi_{y\text{-}corr}$ (equations [6.25] and [6.26]).

$$M_{y\_corr} = E_c I_{c\_corr} \chi_{y\_corr} \qquad [6.25]$$

$$\chi_{y\_corr} = \frac{\frac{f_y}{E_s}}{d - y_{0c\_corr}} \qquad [6.26]$$

The corresponding deflection, v_{y_corr}, may be determined by performing the same calculation as before corrosion by again using equation [6.14], this time replacing the cracked cross-section inertia with the cracked and corroded cross-section inertia (equation [6.27]).

$$v_{y\_corr} = \frac{M_{y\_corr} L^2}{12 E_c I_{c\_corr}} \quad [6.27]$$

6.4.2.2. *Determination of the ultimate moment and the corresponding deflection (corroded beam)*

The ultimate moment of the corroded beam, M_{u_corr}, corresponding to point C_{corr}, is calculated as a cracked cross-section, using a similar calculation to that of the non-corroded beam. The position of the neutral axis, y_{u_corr}, is determined by writing that the normal force is zero, then the moment, M_{u_corr}, as the value resulting from the elementary moments of the stresses around the neutral axis.

Equation [6.28] gives the height of the neutral axis as a non-cracked, corroded cross-section resulting from writing the normal force as equal to zero.

$$y_{u\_corr} = \frac{A_{s\_corr} f_y}{0,8 b f_{cu}} \quad [6.28]$$

The bending moment is then written as resulting from the compression forces in the concrete and tensile forces in the corroded steel (equation [6.29]).

$$M_{u\_corr} = 0,48 y_{u\_corr} b f_{cu}^{\ 2} + A_{s\_corr} f_y (d - y_{u\_corr}) \quad [6.29]$$

The ultimate deflection of the corroded beam, v_{u_corr}, can be calculated by determining whether the rotation capacity of the plastic hinge, which will be written as χ_{pl_corr}, formed after the yielding moment, is limited by the maximum strain allowed under compression of the concrete ε_{cu} or by the maximum allowed extension of the corroded tensile steel reinforcements, ε_{su_corr} (equation [6.30]).

$$\chi_{pl\_corr} = \min\left(\frac{\varepsilon_{cu}}{y_{u\_corr}} ; \frac{\varepsilon_{su\_corr}}{d - y_{u\_corr}} \right) \quad [6.30]$$

The ultimate deflection, v_{u_corr}, is calculated in the same manner as before corrosion (Figure 6.9) and is given by equation [6.31].

$$v_{u\_corr} = v_{y\_corr} + \frac{L}{4}\chi_{pl\_corr} \qquad [6.31]$$

6.4.3. *Numerical application for corroded beam Bs04, compared to a control beam, BsT*

Let us now compare the actual behavior of a non-corroded beam and a corroded beam with the model developed in sections 6.4.1 and 6.4.2. The corroded beam is beam Bs04, for which the corrosion estimation, based on the measurement of cracks induced by corrosion, is presented in Chapter 4. The non-corroded beam is an identical beam, BsT.

The control beam has been tested as of its initial condition, which thus provides access to the complete envelope curve, including the cracking, yielding and ultimate moments (points A, B and C in Figure 6.2).

Corroded beam Bs02 has been tested after being kept under load at a service value greater than 50% with respect to the design at ULS, set out by Eurocode 2. Its loading at failure therefore corresponds to Figure 6.3.

The geometrical data and the characteristics of the materials (steel and concrete), which do not depend on the corrosion intensity, are presented in Table 6.1.

Data	**h** mm	**b** mm	**d** mm	**E_c** **GPa**	**E_s** **GPa**	**f_y** **MPa**	**f_{cu}** **MPa**	**f_{tu}** **MPa**
Beam Bs	280	150	258	30	200	550	45	3

Table 6.1. *Geometric and material data of the BsT beams, independent of corrosion*

The material data influenced by corrosion, steel cross-section and maximum steel elongation at failure, are summarized in Table 6.2.

Data	A_s mm^2	ΔA_s mm^2	A_{s_corr} mm^2	C	ε_{su}	ε_{su_corr}
Beam BsT	226	0	N/A		0.1	N/A
Corroded beam Bs04		54	172	0.24		0.02–0.027

Table 6.2. *Geometric and material data of the BsT beams, dependent on corrosion*

The change in corroded-steel elongation at failure, calculated based on the model by Zhu *et al.* [ZHU 17a], can be seen in Figure 6.11. With a maximum local corrosion rate of 24% of the tensile reinforcements, this leads to a reduction in the elongation at failure comprised between 76% and 82%. This interval will be used to predict the steel-reinforcement failure of beam Bs04.

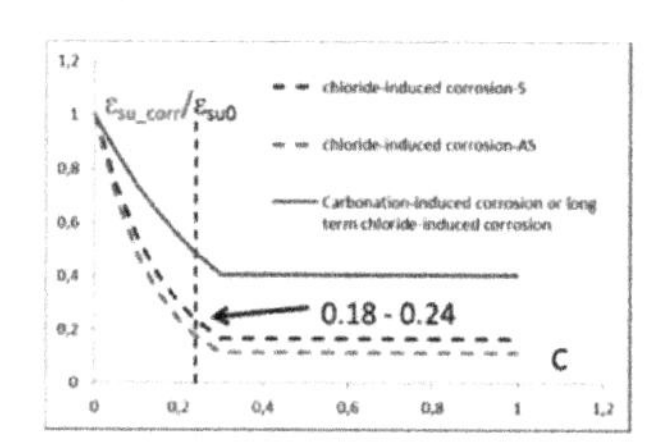

Figure 6.11. *View of the reduction in failure elongation of tensile, corroded steel reinforcements of beam Bs04 between 0.18 and 0.24, according to the model by Zhu* et al. *[ZHU 17a]. For a color version of the figure, see www.iste.co.uk/francois/corrosion.zip*

6.4.3.1. *Comparison of actual behavior and behavior of the model of control beam BsT*

Based on the data of Tables 6.1 and 6.2 it is possible to predict the bending response of control beam BsT. The data of the prediction calculation defined in section 6.4.1 are presented in Table 6.3. This enables the position of the neutral axis in the different behavior stages to be found, along with the cracked and non-cracked inertias and the bending curvatures in the most stressed median cross-section with calculation of the yield curvature corresponding to the failure of the compressed concrete or the tensile steel reinforcements.

Calculation values	y_{0nc} mm	y_{0c} mm	y_u mm	I_{nc} mm^4	I_c mm^4
	144	6.3	2.3	$29.47\ 10^{-5}$	$6.98\ 10^{-5}$
Control beam BsT	χ_{c0} mm^{-1}	χ_y mm^{-1}	χ_{pl} concrete mm^{-1}	χ_{pl} steel mm^{-1}	χ_{pl} min mm^{-1}
	$0.74\ 10^{-3}$	$14.1\ 10^{-3}$	152	426	152

Table 6.3. *Characteristics of the median cross-section of control beam BsT: positions of the neutral axis, inertias and curvatures calculated by the model*

The minimum yield curvature is that corresponding to the compressed-concrete failure value, which is almost three times' lower than that which would lead to failure of the tensile steel reinforcements. Table 6.4 presents the values of the moments corresponding, successively, to the first tensile-concrete cracking, the yielding of the reinforcements and failure, as well as the corresponding deflections.

Calculation values	M_c kN.m	M_y kN.m	M_u kN.m
	6.5	29.5	31
Control beam BsT	v_c mm	v_y mm	v_u mm
	0.5	10	106

Table 6.4. *Bending moments and deflections corresponding to the first tensile-concrete cracking, reinforcement yielding and failure, as calculated by the model*

The comparison of the model's prediction and the actual experimental behavior is presented in Figure 6.12. The failure mode clearly corresponds to the failure of the compressed concrete after formation of a plastic hinge, as can be seen in Figure 6.13. The bending moments and the actual deflections measured experimentally on control beam BsT are presented in Table 6.5. The moment values predicted by the model are very close to those obtained experimentally. The corresponding deflections are also relatively well predicted.

Experimental values	M_c kN.m	M_y kN.m	M_u kN.m
	7	30	32.8
Control beam BsT	v_c mm	v_y mm	v_u mm
	0.8	13	150

Table 6.5. *Bending moments and deflections corresponding to the first tensile-concrete cracking, reinforcement yielding and failure, obtained experimentally for control beam BsT*

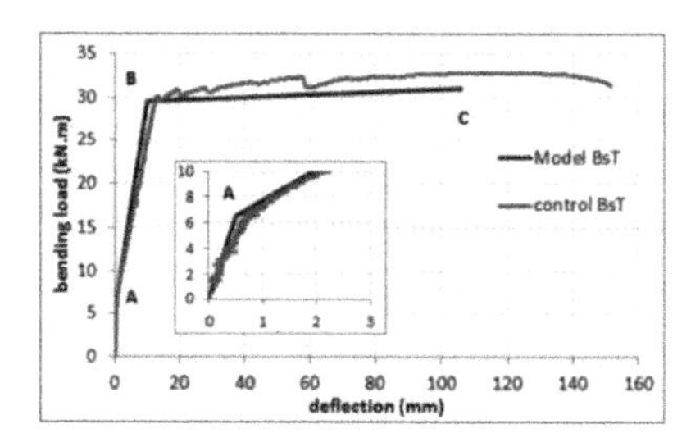

Figure 6.12. *Comparison of the experimental envelope curve and the model developed in section 6.4.1. For a color version of the figure, see www.iste.co.uk/francois/corrosion.zip*

The deflection at failure is underestimated by the model, but the deflection measurement during the course of the test comprises a part further to the failure of the compressed concrete, as long as the tensile reinforcements continue to yield.

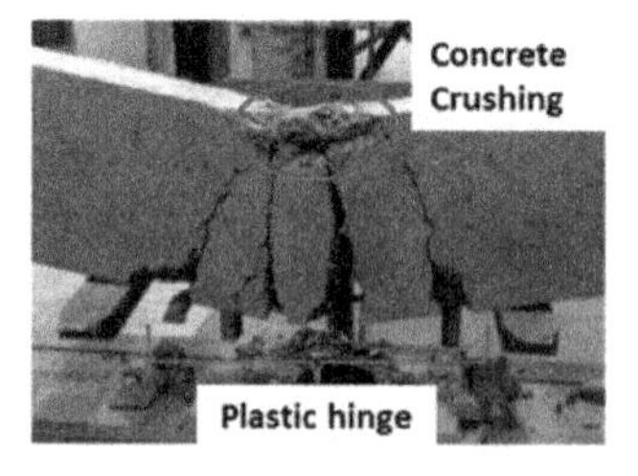

Figure 6.13. *View of the plastic hinge and the compressed-concrete failure for control beam BsT, according to Yu* et al. *[YU 15b]. Note that following the failure of the compressed concrete, an increase in deflection was able to be measured experimentally with the growth in strains in tensile steel reinforcements prior to their failure. For a color version of the figure, see www.iste.co.uk/francois/corrosion.zip*

6.4.3.2. *Comparison of actual behavior and behavior of the model of corroded beam Bs04*

Based on the data of Tables 6.1 and 6.2 it is possible to predict the bending response of corroded beam Bs04. The data of the prediction calculation defined in section 6.4.2 are presented in Table 6.6. Even if beam Bs04 was corroded under a sustained service load, the full set of information is presented for the corroded beam: the position of the neutral axis in the different behavior stages, the cracked and non-cracked inertias and the bending curvatures in the most stressed median cross-section with the calculation of the yield curvature corresponding to the failure of the compressed concrete or the corroded tensile steel reinforcements. Two values of the ratio of failure elongation (0.18 and 0.24) are used as shown in Figure 6.11, which thus correspond to the lower and higher value of the interval in Tables 6.6 and 6.7, and to the mark *L* and *H* on Figures 6.14 and 6.15. Indeed, the envelope curve is a behavior prediction stage that follows corrosion.

Calculation values	y_{0nc} mm	y_{0c} mm	y_u mm	I_{nc} mm^4	I_c
	143	5.6	1.7	$29\ 10^{-5}$	$5.56\ 10^{-5}$
Corroded beam Bs04	χ_{c0} mm^{-1}	χ_y mm^{-1}	χ_{pl} concrete mm^{-1}	χ_{pl} steel mm^{-1}	χ_{pl} min mm^{-1}
	$0.73\ 10^{-3}$	$13.6\ 10^{-3}$	200	75–100	75–100

Table 6.6. *Characteristics of the median cross-section of corroded beam Bs04: positions of the neutral axis, inertias and curvatures. The interval for yield curvature corresponds to the boundaries of corroded steel elongation of 0.18 and 0.24 (Figure 6.11)*

The minimum yield curvature is that corresponding to the corroded tensile reinforcement failure value, which becomes more than twice less that corresponding to the failure of compressed concrete. Table 6.7 presents the values of the moments corresponding, successively, to the first tensile-concrete cracking, the yielding of the reinforcements and failure, as well as the corresponding deflections.

Calculation values	M_c kN.m	M_y kN.m	M_u kN.m
	6.4	22.7	23.8
Corroded beam Bs04	v_c mm	v_y mm	v_u mm
	0.5	9.6	52–70

Table 6.7. *Bending moments and deflections corresponding to the first tensile-concrete cracking, reinforcement yielding and failure, as calculated by the model. The interval for v_u corresponds to the lower and higher boundaries of corroded steel elongation (Figure 6.11)*

The model's prediction is presented in Figure 6.14. The values of the moments and deflections measured experimentally on corroded beam Bs04 are listed in Table 6.8. Only the yielding and ultimate moments are indicated as beam Bs04 was already service cracked before the test. The deflection corresponding to the yielding moment is not indicated as the initial residual deflection of the corroded beam is not known at the time of the test. Indeed, beam Bs04, which was kept under sustained load, was unloaded in order to be able to proceed with the bending test.

Experimental values	M_y kN.m	M_u kN.m
	24.5	28.7
Corroded beam Bs04	v_y mm	v_u mm
	N/A	65

Table 6.8. *Bending moments and deflections corresponding to reinforcement yielding and failure, obtained experimentally for corroded beam Bs04*

As such, the envelope curve calculated with all loading stages (cracking, yielding and failure) needs to be modified to take account of the fact that beam Bs04 was corroded under sustained load at a value of 21.2 kN.m (point D_{corr}). Figure 6.14 provides a schematic representation of the unloading from D_{corr} and the actual behavior of the reloaded beam following this slope. The corresponding curve is presented in Figure 6.15.

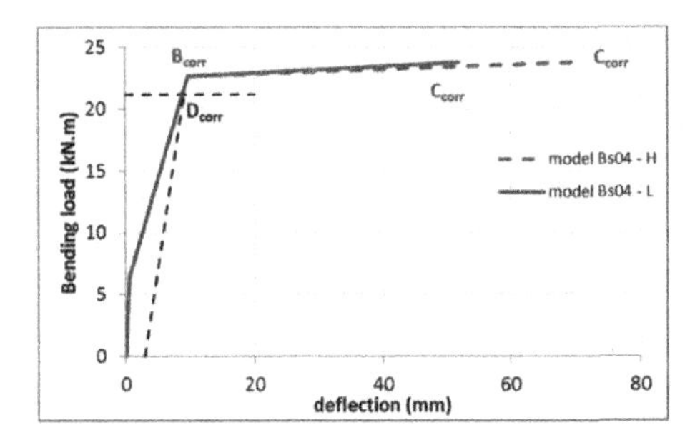

Figure 6.14. *Envelope curve of the corroded beam and updated beam taking account of the historical loading of corroded beam Bs04, based on the model developed in section 6.4.2.* L *and* H *correspond to the lower and higher boundaries of corroded steel elongation according to the Zhu* et al. *model (Figure 6.11). For a color version of the figure, see www.iste.co.uk/francois/corrosion.zip*

It can be observed that for beam Bs04, the service loading (load sustained during the corrosion process) was very high and that the residual bearing capacity calculated by the model is barely higher than this value.

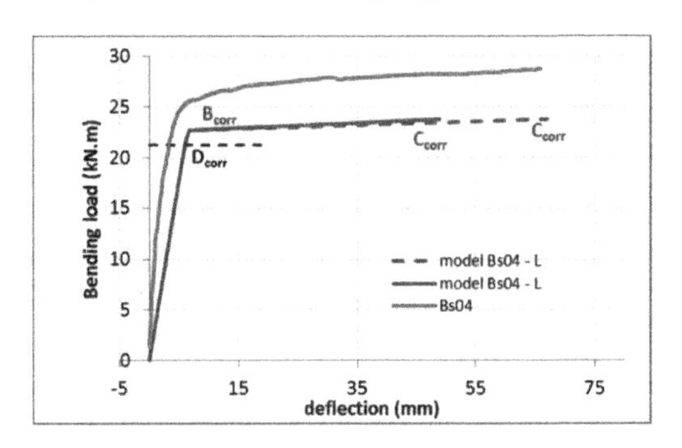

Figure 6.15. *Comparison between the actual behavior of corroded beam Bs04 and the model developed in section 6.4.2.* L *and* H *correspond to the lower and higher boundaries of corroded steel elongation according to the Zhu* et al. *model (Figure 6.11). For a color version of the figure, see www.iste.co.uk/francois/corrosion.zip*

The actual bearing capacity of beam Bs04, 28.7 kN.m, appears higher at the moment calculated by the model, 23.8 kN.m, which corresponds to a deviation of 20%. The actual yielding moment of beam Bs04, 24.5 kN.m, appears higher at the moment calculated by the model, 22.7 kN.m, which corresponds to a deviation of 8%.

The yielding-moment value is closer between the model and the actual behavior because the yield-strength value is identical for steel before and after corrosion. Concerning stress at failure, this differs from the perfect elasto-plastic model with work hardening of the reinforcements [ZHU 14]. Taking this work hardening into account would enable a more accurate prediction.

The failure mode clearly corresponds to the failure of the corroded steel reinforcements after formation of a plastic hinge, as can be seen in Figure 6.16, without failure of the compressed concrete. The deflection at failure calculated by the model planned at between 52 and 70 mm encloses the actual deflection corresponding to a brittle failure mode well.

Figure 6.16. *View of the plastic hinge and the failure of the tensile corroded steel reinforcements for corroded beam Bs04, according to Yu* et al. *[YU 15b]. It should be noted that compressed-concrete failure is not reached as planned in the model. For a color version of the figure, see www.iste.co.uk/francois/corrosion.zip*

6.4.3.3. *Conclusions on the model to predict bearing capacity and deflection at failure*

The model to predict the residual bearing capacity and deflection at failure of a corroded beam, based partly on the conventional reinforced-concrete calculations, but above all on a visual diagnosis of the state of corrosion by measuring the corrosion-induced cracks, gives positive results: deflection and bearing capacity are predicted with a deviation of below 20%, and even below 10% for the yielding moment value.

The actual corrosion state measured further to the campaign of experimental tests demonstrated [YU 15b] that the maximum cross-section loss at the median area of the beam was 24%, which corresponds to the value calculated by the model predicting cross-section losses under localized corrosion (LC) based on the opening of the corrosion cracks, as presented in Chapter 4.

6.5. Assessment of behavior in service (bending stiffness and deflections) of a corroded structure

The prediction of the evolution in a corroded structure's deflections in service is important for the operation of the corroded structure: displacements compatible with its usage in particular. It could also be of interest to know the evolution in bending stiffness with respect to the level of reinforcement corrosion within the framework of a structural diagnosis. Indeed, the measurement of the increase in deflection for an increase in loading performed during the diagnosis could enable, through inverse analysis, the level of the structure's corrosion to be identified, and thus its residual service life to be predicted.

As with the evolution in bearing capacity with respect to the state of corrosion, presented in section 6.4, it is necessary to dispose of a model enabling the deflections and bending stiffness of a non-corroded structure to be planned for in order to then incorporate the corrosion parameter into the model. The deflections of the reinforced-concrete elements are predicted by general regulatory methods, taking into account the supply of tensile concrete between the bending cracks. This is the case for the CEB-FIP Model Code, for example, which used an effective tensile member cross-section [CEB 99].

To calculate the behavior in service of reinforced concrete, the stiffening effect of tensile concrete needs to be accurately modeled. This could be the case of a calculation by finite elements by modifying the constitutive relation of tensile concrete, taking into account the softening of the post-peak stress [BAZ 80, GUP 89, VEC 00]. But in reality, the tension stiffening is due to the steel-concrete bond between bending cracks. Thus, another approach consists of using bond models that are based on a distribution of the bond stress along the reinforcement, respecting the equilibria in force and the compatibility of the strains [SOM 81, FLO 82, CHA 92]. This approach, even adopted in a finite-element formulation, presents certain limits, as it requires an assumption as to the bond stress distribution along the reinforcement. Other models, which emphasize the calculation of global behavior, are based on analytical, empirical moment-curvature relationships [CAR 86, BIS 05, ALW 90, PRA 90]. The calculation is then substantially simplified with respect to previous methods and the resulting deflection prediction is acceptable but not very accurate.

The method chosen in this chapter to take into account the stiffening effect of tensile concrete is to homogenize the behavior of the beam area situated between two consecutive bending cracks, which will thus be considered as a finite macro-element (ME) of length L_{elem}, to deduce an average inertia (see Figure 5.12). It will be possible to express this average inertia with respect to the corrosion of the tensile reinforcements by the damage variable of the steel-concrete bond defined in Chapter 5. The assembly of the MEs (seen in Figure 6.17) constituting a beam (or, more generally, a concrete structure), will enable the deflections under bending load to be calculated.

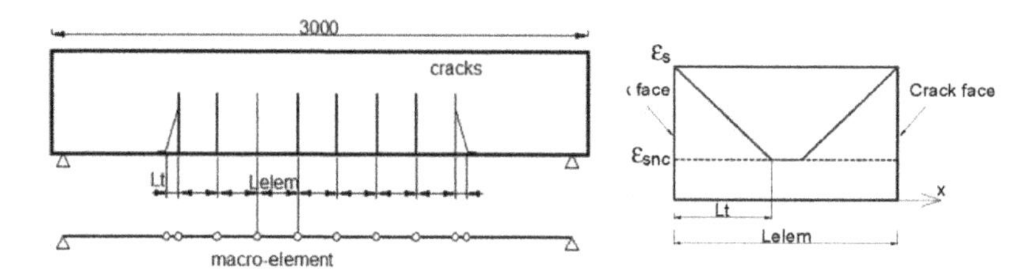

Figure 6.17. *View of a Finite Macro Element between two bending cracks and assembly of the MEs in order to model a reinforced-concrete beam, according to François* et al. *[FRA 06b]. For a color version of the figure, see www.iste.co.uk/francois/corrosion.zip*

It will be assumed for the calculations that follow that the mechanical behavior in the ME is symmetrical, which comes down to writing that the bending moment is constant on the spacing between two consecutive cracks. Therefore, for the continuation of the calculation we will reason based on half of the ME.

6.5.1. *Bending stiffness before corrosion*

The bending stiffness before corrosion of a reinforced-concrete element is directly linked to the average inertia of the beam ME located between the service bending cracks. As indicated in section 5.3.1, and repeated in Figure 6.18, there are two scenarios to be taken into consideration in order to formulate the ME: firstly, the case where the spacing between cracks is more than twice the transfer length (crack formation phase), and secondly, the case where the spacing between cracks is less than twice the transfer length (cracking stabilization phase).

The calculation of this average inertia is based on the assumption that between two bending cracks, there is a linear variation of the position of the neutral axis and the stress of the tensile steel reinforcements along the transfer length. Below, the calculation will be performed for both cracking phases: formation and stabilization.

6.5.1.1. *Average inertia of the Macro-Element in the service-crack formation phase*

In the crack formation phase the transfer length is smaller than the Macro-Element half-length: $L_t \leq \frac{L_{elem}}{2}$, this scenario can be viewed in Figure 6.18. The linear strain variation leads to equations [6.32] and [6.33], as a function of x and that of the neutral-axis position leads to equations [6.34] and [6.35], as a function of x.

$$\varepsilon_s(x) = \left(\frac{\varepsilon_{snc} - \varepsilon_s}{L_t} \right) x + \varepsilon_s \qquad [6.32]$$

For $0 \leq x \leq L_t$ and

$$\varepsilon_s(x) = \varepsilon_s \qquad [6.33]$$

for $L_t \leq x \leq \frac{L_{elem}}{2}$

$$y_0(x) = \left(\frac{y_{0nc} - y_{0c}}{L_t} \right) x + y_{0c} \qquad [6.34]$$

For $0 \leq x \leq L_t$ and

$$y_0(x) = y_{0c} \qquad [6.35]$$

for $L_t \leq x \leq \frac{L_{elem}}{2}$

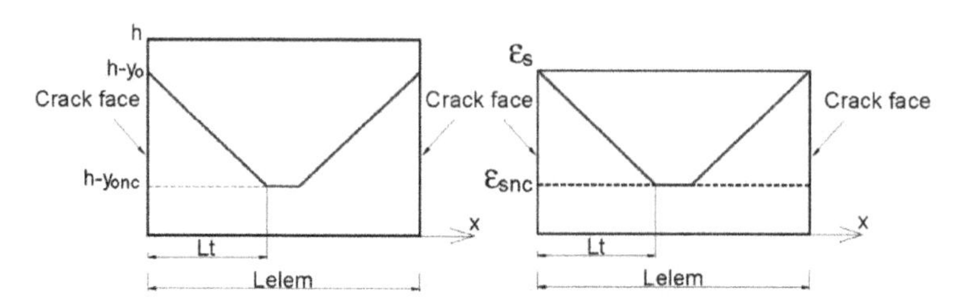

Figure 6.18. *Linearity assumptions concerning neutral-axis position and strain variation in tensile steel on the Finite Macro-Element, according to François* et al. *[FRA 06b]. For a color version of the figure, see www.iste.co.uk/francois/corrosion.zip*

The curvature is calculated along the Macro-Element then the average curvature is deduced by homogenization, which then enables the average inertia to be calculated. The bending curvature at all points along half of the Macro-Element is then written according to equations [6.36] and [6.37], as a function of x.

$$\chi(x) = \frac{\varepsilon_s(x)}{d - y_{0c}} \qquad [6.36]$$

For $0 \leq x \leq L_t$ and

$$\chi(x) = \chi_{nc} \qquad [6.37]$$

for $L_t \leq x \leq \frac{L_{elem}}{2}$

We can then determine the average curvature, χ_a, through homogenization along the length of the Macro-Element (equation [6.38]).

$$\frac{L_{elem}}{2}\chi_a = \int_0^{L_t} \chi(x)dx + \int_{L_t}^{\frac{L_{elem}}{2}} \chi_{nc} dx \qquad [6.38]$$

The average inertia, I_a, of the average curvature value is deduced by writing the moment-curvature relation (equation [6.39])

$$I_a = \frac{M}{E_c \chi_a} \qquad [6.39]$$

By developing equation [6.38] to calculate the average curvature and using equation [6.39], an expression of the average inertia is obtained, which is written according to equation [6.40],

$$I_a = \frac{L_{elem}}{2}\left[\left(\frac{d_{nc}}{I_{nc}} - \frac{d_c}{I_c}\right)\frac{L_t}{y} + \left(\frac{d_c}{I_c} - \frac{d_c d_{nc}}{I_{nc} y} + \frac{d_c^2}{I_c y}\right)\frac{L_t}{y}\ln\left(\frac{d_{nc}}{d_c}\right) + \frac{1}{I_{nc}}\left(\frac{L_{elem}}{2} - L_t\right)\right]^{-1} \qquad [6.40]$$

whereby $d_c = d - y_{0c}$, $d_{nc} = d - y_{0nc}$, $y = y_{0c} - y_{0nc}$

6.5.1.2. *Average inertia of the Macro-Element in the service-crack stabilization phase*

In the crack stabilization phase the transfer length is larger than the Macro-Element half-length: $L_t \geq \frac{L_{elem}}{2}$.

The previous calculation (section 6.5.1.1) is barely changed, with the expression of the average curvature written according to equation [6.41].

$$\frac{L_{elem}}{2}\chi_a = \int_0^{\frac{L_{elem}}{2}} \chi(x)dx \qquad [6.41]$$

This enables calculation of the average inertia, I_a (equation [6.42]), by combining with equation [6.39],

$$I_a = \frac{L_{elem}}{2}\left[\left(\frac{d_{nc}}{I_{nc}} - \frac{d_c}{I_c}\right)\frac{L_{elem}}{2y} + \left(\frac{d_c}{I_c} - \frac{d_c d_{nc}}{I_{nc} y} + \frac{d_c^2}{I_c y}\right)\frac{L_t}{y}\ln\left(\frac{y}{d_c}\frac{L_{elem}}{2L_t} + 1\right)\right]^{-1} \qquad [6.42]$$

whereby $d_c = d - y_{0c}$, $d_{nc} = d - y_{0nc}$, $y = y_{0c} - y_{0nc}$

6.5.1.3. *Calculating the deflection of a non-corroded cracked beam*

The average inertia, I_a, obtained takes account of the stiffening effect of the tensile concrete via the transfer length. This parameter was assessed by François *et al.* [FRA 06b] and appears of the order of 100 mm for a deformed reinforcement. Except when conducting a campaign of specific tests, it is recommended to take this value of 100 mm. Each Macro-Element, *i*, is allocated its average inertia, $I_a(i)$, to perform a conventional calculation of beam structures (Figure 6.19). If the spacing between cracks is constant, the entirety of the cracked length can be replaced by a segment of beam presenting an inertia, I_a, as seen in Figure 6.19.

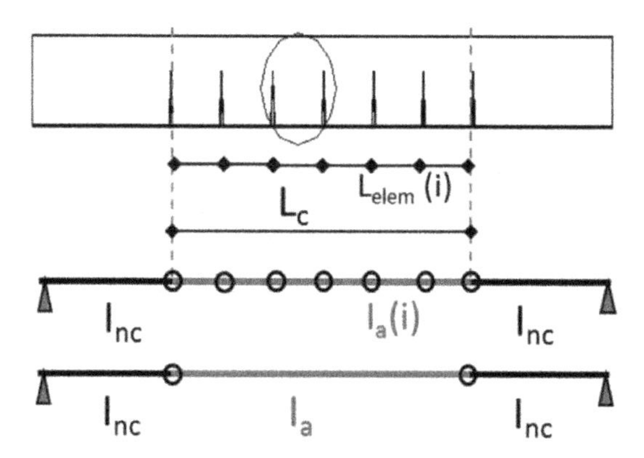

Figure 6.19. *Assembly of Macro-Elements to calculate the deflection of a non-corroded beam. If the spacing is constant (the general case), the average inertia may be used along the entire cracked length. For a color version of the figure, see www.iste.co.uk/francois/corrosion.zip*

6.5.2. *Bending stiffness after corrosion*

The bending stiffness after corrosion of a reinforced-concrete element will depend on the corrosion, taking account of the bond reduction due to corrosion on each Macro-Element by means of the damage variable, D_c, defined in section 5.3.2. The main cause of bond reduction is the creation of corrosion cracks, which leads to a deconfinement of the reinforcements. Nevertheless, the presence of shear stirrups strongly contributes to maintaining a significant bond on the deformed bars, hence Fang *et al.* [FAN 04] concluded that there is no bond loss due to corrosion in the case of a confinement. Unfortunately, this study is based on accelerated corrosion.

Yet, accelerated corrosion, whilst not comparable to natural corrosion, leads to the creation of cracks parallel to the reinforcements, the effect of which on bonding is influenced by the presence of stirrups. Likewise, the forces at the supports contribute to significant confinement of the corroded reinforcements in the concrete being maintained, and limit the risks of anchorage losses [LUN 15].

The consideration of corrosion on the bond between corroded steel and concrete is, as indicated in Chapter 5, based on an increase in the transfer length. As this increase is due, in phenomenological terms, to the creation of corrosion cracks, it is therefore quantified by the localized corrosion (LC). As this is the first stage in corrosion, it leads to the creation of corrosion cracks. The damage variable, D_c, defined in section 5.3.2, will therefore be based on local corrosion (LC).

As corrosion does not necessarily affect the entirety of a structure or of a structural element, it is necessary to quantify the bond loss of each bar. Indeed, the presence of non-corroded bars leads to perfect bond, which would be poorly taken into account if reasoning in terms of the entire reinforcement cross-section of the reinforced concrete cross-section. Figure 6.20 illustrates two different corrosion states for a cross-section of beam composed of three tensile bars.

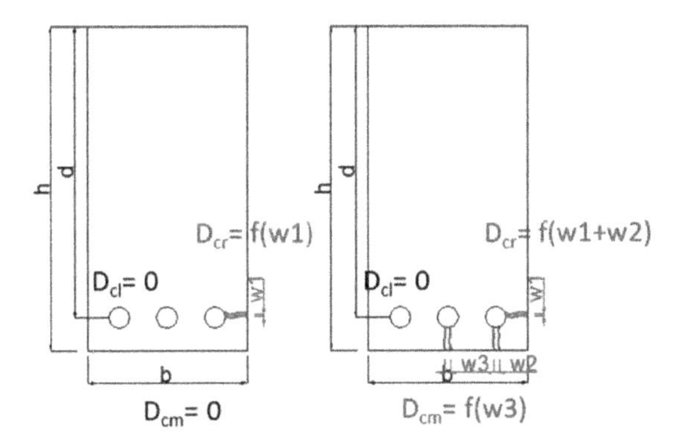

Figure 6.20. *Consideration of the bond damage, bar by bar, D_{ci}, D_{cm}, D_{cr}, with respect to the corrosion cracks visible on the concrete surface. For a color version of the figure, see www.iste.co.uk/francois/corrosion.zip*

In the case of the beam presented in Figure 6.20, there are therefore three damage variables to be calculated: one for each reinforcement. From left to right, D_{cl}, D_{cm}, and D_{cr} are obtained, which depend on the presence of corrosion cracks that may be attributed to the corrosion of the three corresponding bars.

The average bond damage, D_{ca}, is then calculated as the average damage across all of the bars (here: three), according to equation [6.43].

$$D_{ca} = \frac{D_{cl} + D_{cm} + D_{cr}}{3} \quad [6.43]$$

The new transfer length is then deduced following equation [6.44].

$$L_{tcor} = \frac{L_t}{1 - D_{ca}} \quad [6.44]$$

As in section 6.5.1, it would be necessary to separate the calculation of average inertia after corrosion according to whether the current phase is that of service crack formation or that of stabilized service cracking, which depends on the ratio between crack spacing and transfer length. However, given the fact that the transfer length is increased by corrosion $L_{tcor} > L_t$, we find ourselves almost systematically in the scenario corresponding to stabilized cracking after corrosion.

The average inertia of each beam ME between two consecutive service cracks may be recalculated by taking account of the new transfer length. During a structure's corrosion diagnosis, we do not know whether the current phase is the essentially localized corrosion phase or whether the corrosion has begun to generalize along the cracks induced by corrosion. We therefore suggest enclosing the possible variation in bending stiffness due to corrosion with a high value corresponding to solely taking into account the change in transfer length in the calculation of average inertias of the Macro-Element (equation [6.45]) and a low value corresponding to also taking into account the generalized corrosion (GC) on the Macro-Element (equation [6.46]). Only equations [6.45] and [6.46] corresponding to the case where the transfer length is higher than the half-length of the Macro-Element are presented (rewriting of equation [6.42]). Indeed, this is the most probable scenario given that the corrosion increases transfer length, $L_{tcor} \geq \frac{L_{elem}}{2}$. Where this is not the case, equation [6.40] simply needs to be rewritten following the same principle.

Equation [6.45] corresponds to simply changing the transfer length based on the local corrosion (LC), with all other variables equal to those calculated prior to corrosion.

$$I_{a\_corr} = \frac{L_{elem}}{2}\left[\left(\frac{d_{nc}}{I_{nc}} - \frac{d_c}{I_c}\right)\frac{L_{elem}}{2y} + \left(\frac{d_c}{I_c} - \frac{d_c d_{nc}}{I_{nc} y} + \frac{d_c^2}{I_c y}\right)\frac{L_t}{y}\ln\left(\frac{y}{d_c}\frac{L_{elem}}{2L_{tcor}} + 1\right)\right]^{-1} \quad [6.45]$$

whereby $d_c = d - y_{0c}$, $d_{nc} = d - y_{0nc}$, $y = y_{0c} - y_{0nc}$

Equation [6.46] corresponds to the change in transfer length based on the local corrosion (LC) and the change in all other variables relating to the characteristics of the only cracked cross-section, recalculated to take account of the generalized corrosion (GC). Indeed, given the low impact of the cross-section of reinforcements on the characteristics of the non-cracked cross-section (position of the neutral axis, non-cracked inertia), we propose to retain their initial values, prior to corrosion, when calculating the stiffness of the corroded element.

$$I_{a\_corrG} = \frac{L_{elem}}{2}\left[\left(\frac{d_{nc}}{I_{nc}} - \frac{d_{c\_corG}}{I_{c\_corG}}\right)\frac{L_{elem}}{2y_{corG}} + \cdots\right]^{-1} \quad [6.46]$$

$$\cdots = \left(\frac{d_{c\_corG}}{I_{c\_corG}} - \frac{d_{c\_corG} d_{nc}}{I_{nc} y_{corG}} + \frac{d^2_{c\_corG}}{I_{c\_corG} y_{corG}}\right)\frac{L_{tcor}}{y_{corG}}\ln\left(\frac{y_{corG}}{d_{c\_corG}}\frac{L_{elem}}{2L_{tcor}} + 1\right)$$

whereby $d_{c\_corG} = d - y_{0c\_corG}$, $d_{nc} = d - y_{0nc}$, $y_{corG} = y_{0c\_corG} - y_{0nc}$

with y_{0c_corG} the position of the neutral axis on the cracked cross-section, calculated using the average corroded cross-section (GC).

To complete the calculation of bending stiffness after corrosion, each Macro-Element, *i*, needs to be allocated its new average inertia, $I_{a\_corr}(i)$ or

$I_{a\_corrG}(i)$, calculated after corrosion for both scenarios, respectively: local corrosion (LC) phase or generalized corrosion (GC) phase, in order to carry out a conventional calculation of beam structures (Figure 6.21). In the case where the corrosion also affects areas of non-service-cracked structures, there is no inertia change taken into consideration for calculation in the local corrosion (LC) phase, nor in the generalized corrosion (GC) phase, owing to the low impact of the reinforcements on the non-cracked characteristics.

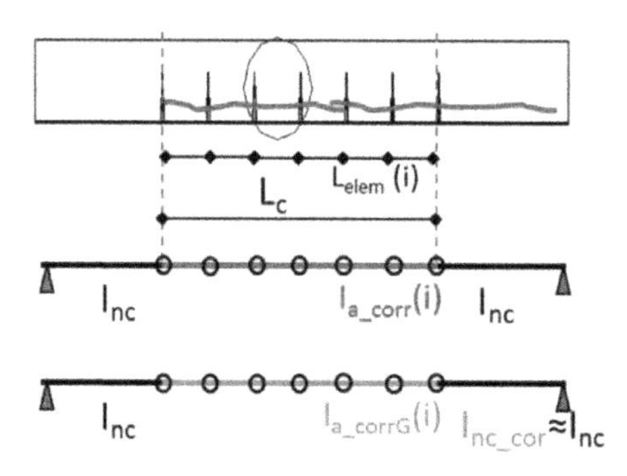

Figure 6.21. *Assembly of Macro-Elements to calculate the deflection of a corroded beam. In the presence of corrosion in areas without service cracks, its effect on the change in inertia is disregarded. For a color version of the figure, see www.iste.co.uk/francois/corrosion.zip*

6.5.3. *Numerical application for corroded beam Bs04, compared to a control beam, BsT*

Let us now compare the actual behavior of a non-corroded beam and a corroded beam with the model developed in sections 6.5.1 and 6.5.2. The corroded beam is beam Bs04, for which the corrosion estimation, based on the measurement of cracks induced by corrosion, is presented in Chapter 4. The non-corroded beam is an identical beam, BsT.

As with the comparison of the model predicting bearing capacity with respect to the state of corrosion, it is not possible to have access to exactly the same predictions on the control beam, which has been tested without having been loaded beforehand, and the corroded beam, which spent several years kept under a significant service load.

The geometric data of control beam BsT and beam Bs04 are already given in Tables 6.1 and 6.2 above.

6.5.3.1. *Comparison of actual bending stiffness and bending stiffness of the model of control beam BsT*

Based on the data of Tables 6.1 and 6.2 it is possible to predict the bending response of control beam BsT by adding the calculation of average inertia, I_a, for the cracked area. The spacing between service cracks measured on control beam BsT is L_{elem} = 200 mm (corresponding to the spacing between the shear stirrups). This value corresponds to double the value of the usual transfer length, L_t = 100 mm, which leads to a single cracking phase for this beam.

The values calculated are presented in Table 6.9.

Calculation values	y_{0nc} mm	y_{0c} mm	y mm	I_{nc} mm^4	I_c mm^4
	144	6.3	–81.4	29.47 10^{-5}	6.98 10^{-5}
Control beam BsT	L_{elem} mm	L_t mm	I_a mm^4	d_c mm	d_{nc} mm
	200	100	10.2 10^{-5}	195.3	113.9

Table 6.9. *Characteristics of the Macro-Element of control beam Bs, calculated by the model*

The assembly of the Macro-Elements corresponding to Figure 6.19 enables the bending stiffness of the control beam to be calculated, and therefore the slope of unloading then reloading during load-unload cycles.

The bending stiffness value can be approximated by assuming that the whole of the beam has a constant inertia equal to the average inertia (see Figure 6.22); indeed, the cracked length, L_c, calculated with the cracking and yielding moment values (Table 6.4) has a value of L_c = 2.2 m. The relative deviation between a calculation with a constant inertia across the whole length of the beam or an inertia value equaling the average inertia along the cracked length and the value prior to cracking on the edges is below 1%.

Quite simply, the stiffness, *k*, is then obtained according to equation [6.47].

$$k = \frac{12E_c I_a}{L^2} = 4670kN \quad [6.47]$$

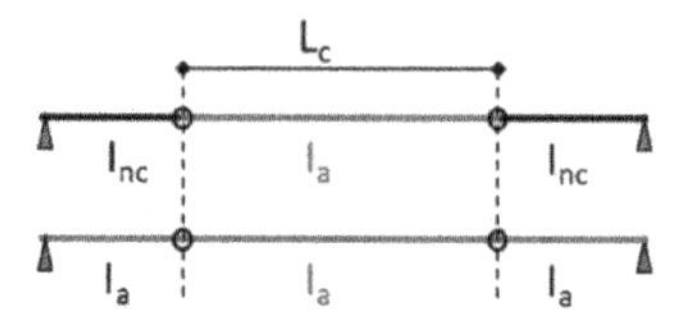

Figure 6.22. *Approximation of a constant inertia in order to calculate the bending stiffness of service-cracked control beam BsT. For a color version of the figure, see www.iste.co.uk/francois/corrosion.zip*

The comparison between the stiffness calculated and the value measured experimentally is presented in Figure 6.23. We proceeded with a loading-unloading cycle beyond the cracking load to obtain the stiffness value of the cracked beam. Given that this beam, BsT, has only one cracking phase, i.e. the formation phase corresponds to the crack stabilization phase, the bending stiffness remains the same regardless of the loading achieved (point D on Figure 6.23).

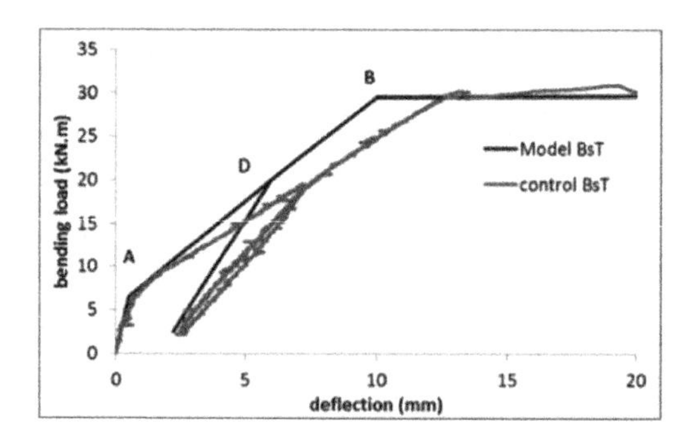

Figure 6.23. *View of the bending stiffness of control beam BsT, beyond its cracking load: comparison of the model and the experimental value. For a color version of the figure, see www.iste.co.uk/francois/corrosion.zip*

It can be observed that the model's prediction slightly overestimates the bending stiffness after cracking. It should be noted that slope AB corresponding to the cracking phase was also overestimated.

6.5.3.2. *Comparison of actual bending stiffness and bending stiffness of the model of corroded beam B04*

In order to predict the stiffness of the corroded beam, it is necessary to determine the damage variable, D_c, of the bond for each tensile reinforcement and for each ME located between two consecutive service cracks.

The damage variables are calculated based on the localized corrosion (LC) expressed as an absolute value, ΔA_s, for each interval between two cracks and for a spacing of 200 mm for non-cracked areas (corresponding to the length of a Macro-Element). The corresponding values are presented in Table 6.10.

The damage variable, D_c, for a given bar is calculated by equation [6.48] (section 5.3.2). This variable depends on the existence of a confinement brought by the shear stirrups, which is the case for beam Bs04. Each tensile reinforcement is 12 mm in diameter, therefore the cross-section of a bar is calculated as $As = 113$ mm^2. The shear stirrups are of diameter 6 mm and are spaced 200 mm apart. It is therefore considered that for each Macro-Element, each tensile bar is maintained by a ϕ of 6 mm at each end $Ast = 56.5$ mm^2. This enables the value of st to be obtained, resulting from equation [5.6] and calculated with equation [6.48].

$$st = 1 - 1{,}8\frac{A_{st}}{A_s}\left(1 - 0.64\frac{A_{st}}{A_s}\right) = 0.64 \qquad [6.48]$$

The calculation of the damage variable for each Macro-Element thus results from equation [6.49], obtained by writing $pf_g = 4$ (localized corrosion) and $st = 0.64$ in equation [5.5].

$$D_c = \min\left(0{,}32\log(1 + 0.25(\Delta A_s - \Delta A_{s0});\ 1\right) \qquad [6.49]$$

To calculate the damage variable for each Macro-Element of length 200 mm, the maximum local-corrosion (LC) value is used on the two half intervals of 100 mm, corresponding to the position of the Macro-Element based on Table 4.2. The transfer length after corrosion, L_{tcor}, can then be calculated. Using equation [6.45] it is then possible to obtain the average inertia of each Macro-Element based on the characteristics of the beam before corrosion, except of course the consideration of the change in transfer length.

Interval (mm)	ΔA_{sf} mm^2	ΔA_{sb} mm^2	D_{cf}	D_{cb}	D_{ca}	L_{tcor} mm	I_{a_corr} m_4
0–200	28.7	3.7	0.28	0.02	0.15	118	9.31E-05
200–400	27.7	3.8	0.27	0.03	0.15	118	9.31E-05
400–600	12.4	4.2	0.17	0.04	0.105	112	9.54E-05
600–800	13.8	4.6	0.18	0.05	0.115	113	9.49E-05
800–1,000	14.7	9.5	0.19	0.13	0.16	119	9.27E-05
1,000–1,200	24.4	18.7	0.26	0.22	0.24	132	8.90E-05
1,200–1,400	32.3	21.7	0.29	0.24	0.265	136	8.81E-05
1,400–1,600	35.7	15.2	0.31	0.19	0.25	133	8.88E-05
1,600–1,800	30.5	14.8	0.29	0.19	0.24	132	8.90E-05
1,800–2,000	19.8	12.6	0.23	0.17	0.2	125	9.08E-05
2,000–2,200	13.4	19.9	0.18	0.23	0.205	126	9.06E-05
2,200–2,400	9.6	14	0.14	0.18	0.16	119	9.27E-05
2,400–2,600	22	7.1	0.24	0.1	0.17	120	9.24E-05
2,600–2,800	11.9	15.7	0.16	0.2	0.18	122	9.17E-05
2,800–3,000	8.5	9.8	0.12	0.14	0.13	115	9.42E-05

Table 6.10. *Calculation of the bond damage variable for each Macro-Element of beam Bs04 and the corresponding values of the transfer length and average corroded inertia (localized corrosion (LC) model)*

The corroded inertia, I_{a_corr}, of each Macro-Element calculated in Table 6.10 enables the bending stiffness of corroded beam Bs04 to be calculated under the scenario of localized corrosion, i.e. in the case where the corrosion has modified the bond owing to the presence of corrosion-induced cracks, but where there is no reduction in the bar cross-sections all along the beam.

As explored in section 6.1.5.3, it is necessary to also perform a calculation of the corroded beam's stiffness, by applying the scenario where the corrosion has generalized along the corrosion-induced cracks and has therefore lead to a loss in generalized cross-section (GC) along the length of each Macro-Element. In this case the characteristics of the beam need to be recalculated for each Macro-Element, in particular the position of the neutral axis in a cracked cross-section and the cracked inertia, after modifying the steel cross-section that becomes that calculated after application of the generalized-corrosion (GC) calculation model. As indicated in section 6.5.2, it is not necessary to modify the characteristics of the non-cracked cross-section (position of the neutral axis and non-cracked inertia) as the reinforcement cross-section has a very low influence.

Interval (mm)	D_{ca}	L_{tcor} mm	A_{scorG} mm_2	y_{oc_corG} mm	I_{c_corG} mm^4	I_{a_corrG} m_4
0–200	0.15	118	217	6.16	6.75E-05	9.02E-05
200–400	0.15	118	217	6.16	6.76E-05	9.03E-05
400–600	0.105	112	224	6.24	6.92E-05	9.46E-05
600–800	0.115	113	223	6.23	6.90E-05	9.39E-05
800–1,000	0.16	119	226	6.26	6.98E-05	9.27E-05
1,000–1,200	0.24	132	218	6.17	6.77E-05	8.64E-05
1,200–1,400	0.265	136	209	6.06	6.55E-05	8.29E-05
1,400–1,600	0.25	133	211	6.08	6.59E-05	8.40E-05
1,600–1,800	0.24	132	213	6.11	6.65E-05	8.50E-05
1,800–2,000	0.2	125	218	6.17	6.77E-05	8.82E-05
2,000–2,200	0.205	126	218	6.16	6.76E-05	8.78E-05
2,200–2,400	0.16	119	221	6.21	6.86E-05	9.12E-05
2,400–2,600	0.17	120	219	6.18	6.81E-05	9.02E-05
2,600–2,800	0.18	122	220	6.19	6.82E-05	8.97E-05
2,800–3,000	0.13	115	224	6.2	6.92E-05	9.33E-05

Table 6.11. *Calculation of the corroded inertia, taking into account the generalized corrosion (GC) for each Macro-Element of beam Bs04*

It should be noted that for beam Bs04, the generalized corrosion is low ($\leq$7.5%), the corroded inertias with the local-corrosion (LC) scenario (Table 7.10) or the generalized-corrosion (GC) scenario (Table 6.11) are barely different with a relative deviation in the vicinity of 6%.

Calculating the corresponding bending stiffness requires the method of finite elements to be used in a “beam” formulation, by assembling the different Macro-Elements, which each have a different corroded inertia.

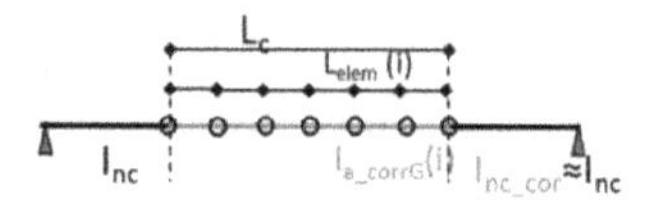

Figure 6.24. *Assembly of the Macro-Elements to calculate bending stiffness of corroded beam Bs04. For a color version of the figure, see www.iste.co.uk/francois/corrosion.zip*

The comparison between the stiffness calculated and the value measured experimentally is presented in Figure 6.25. We proceeded with a loading-unloading cycle beyond the historical load (that corresponding to the loading sustained during the corrosion phase) to obtain the stiffness value of the corroded beam. With the actual yielding load being close to the historical load, the unloading-reloading cycle took place beyond the yielding load; nevertheless, the stiffness measured corresponds to the bending stiffness of the corroded beam.

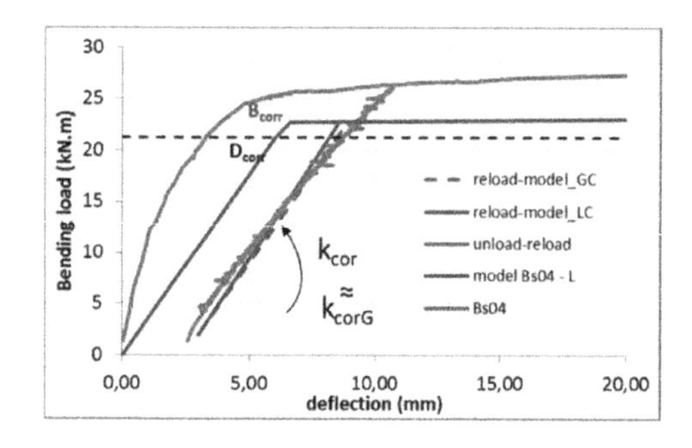

Figure 6.25. *View of the bending stiffness of corroded beam Bs04, beyond its yielding load: comparison of the model and the experimental value. For a color version of the figure, see www.iste.co.uk/francois/corrosion.zip*

We expected the initial slope during loading of beam Bs04 to be also be identical to that of the unloading-reloading cycle; it is probable that the time period between the end of the corrosion test and therefore the suppression of sustained loading, and the bending test, led to stiffening through a phenomenon involving the sticking of the service crack surfaces.

6.6. Conclusion

The prediction of behavior at ULS and at SLS of the corroded elements is possible based on the local (LC) and generalized (GC) cross-section losses assessed by the non-destructive diagnosis resulting from the recording of the corrosion-induced cracks.

NOTE.– If the bearing capacity is well correlated with the local (LC) maximum loss in the most stressed cross-sections, particular attention needs to be paid to the change in ductility induced by corrosion, which is not proportional to the corrosion intensity but is significant from the start of the corrosion process.

The recalculation of displacements is more complex as it requires consideration of the reinforcement confinement brought by the shear stirrups and the structural supports.

7

Predicting the Service Life of Structures

7.1. Introduction

The aim of this chapter is to present the possible predictions concerning the different stages of aging in structures likely to corrode: corrosion initiation, corrosion propagation, prediction of the occurrence of corrosion cracks, service limit state (maximum openings of corrosion-induced cracks, spalling), ultimate limit state (loss of bearing capacity, loss of ductility). For this, the four-phase phenomenological model of the corrosion process is used [FRA 94a] (Figure 7.1), which applies to all structures.

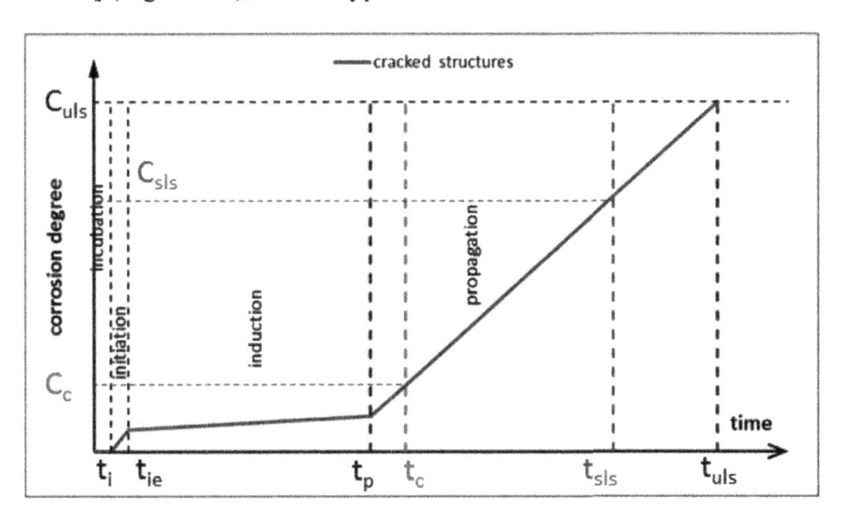

Figure 7.1. *Phenomenological description of the four-phase corrosion process (François* et al. *[FRA 94a]). For a color version of the figure, see www.iste.co.uk/francois/corrosion.zip*

In the corrosion process, the corrosion initiation duration at service crack tips is known as t_i. The duration, t_i, is short in the presence of service cracks: a few days, therefore negligible with respect to the structure's service life. t_{ie} is used to refer to the time corresponding to the formation of a rust "plug" at the crack tip, which substantially slows down the development of corrosion [GHA 17b]. This duration is also short and is negligible with respect to the structure's service life. The start of the propagation phase is annotated t_p, defined as the arrival of the carbonation front at the reinforcements or the reaching of a critical chloride threshold, a debatable concept [ANG 17a]. Much too often this value, t_p, of the start of the propagation phase is used as the end of the structure's service life, yet the safe operation of the structure after the start of corrosion can continue significantly beyond t_p [ZHU 16a, ZHU 16b]. Next, t_c is used to denote the duration corresponding to the creation of the first corrosion-induced cracks; a local corrosion rate, annotated C_c, is allocated to this duration. Thus, this duration value corresponds to the first visible signs of corrosion development. Pursuing the corrosion phase leads to the successive reaching of service limit criteria of the structure, such the maximum opening of the corrosion cracks, the risk of spalling and the increase in deflections, associated with a time, t_{sls} and a corrosion, C_{sls}. Lastly, we arrive at the end of service life in terms of an insufficient bearing capacity, or risk of brittle fracture, with a duration, t_{uls} and a local rate of corrosion, C_{uls}.

7.2. Corrosion initiation

The phenomenon of reinforcement corrosion is often still described in two phases: initiation and propagation. In this case, with respect to the four-phase model of Figure 7.2, the initiation phase corresponds to the first three phases prior to the corrosion propagation phase starting. In these stages before propagation, corrosion is said to be passive, and in the propagation phase the corrosion is said to be active.

This separation into two stages implicitly leads to defining a bifurcation of the equilibrium between a non-corroded state and a corroded state. This perception is inexact as the steel in concrete corrodes even in the passive state, but in passive state the corrosion rate is extremely low and negligible for the service life of Civil Engineering structures (max. 100 years). Nasser *et al.* [NAS 10] measured, for example, corrosion current densities in the passive state that were below 0.05 μA/cm^2, corresponding to a uniform corrosion thickness of 0.05 μm/year (case of passive steel).

The passage from passive corrosion state to an active state (with a significantly higher kinetics) is known as the "depassivation" of the reinforcements. Two main depassivation mechanisms are identified for reinforcements of concrete structures (see Chapter 1): the decrease in pH of the concrete interstitial solution linked to the penetration of atmospheric CO_2 into the porosity of the concrete, and local damage to the passive layer in the presence of chlorides, the exact mechanism of which remains as yet unknown. While the action of carbon dioxide leads to *global depassivation* as there is indeed a destabilization of the passive layer due to the decrease in pH, the mechanism is not the same for the chloride action as the oxides and hydroxides present in the passive layer are not dissolved in the presence of chlorides [TRA 16]; here there are local failures in the passive film, which can nevertheless be referred to as local depassivation.

7.3. Corrosion propagation

The non-uniform nature of corrosion in reinforced concrete leads to a corrosion propagation that depends on several processes: anodic, cathodic and ohmic. The anodic process is influenced by defects in the passive layer and also depends on the growth of the layer of products formed; the cathodic process mainly depends on the dioxygen availability. It may be limited by the diffusion of dioxygen within the concrete. The ohmic process of ion mobility within the interstitial solution depends on the concrete resistivity. For the macro-cell corrosion process (localized corrosion), the geometric effects need to be added; surfaces of the anodic sites, surfaces of the cathodic sites, ratio between anodic and cathodic sites and distances between anodic and cathodic sites (Chapter 1).

The corrosion kinetics thus results from the resolution of a complex system that may be expressed using the Butler-Volmer relation (Chapter 1) for each anodic and cathodic process but that requires knowledge of the geometry of the anodic and cathodic areas as well as the resistivity of the electrolytic environment (the concrete).

Nevertheless, values for corrosion kinetics can be found in the literature (or for corrosion current density, expressed in μA/cm^2), which class the level of aggressiveness of the environmental surroundings and enable an estimation of a structure's service life. These corrosion kinetics are also known as corrosion propagation rate and quantified in terms of μA/cm² of anodic surface area, or thickness loss, expressed in mm/year. The obtaining of these corrosion kinetics is based on the use of Faraday's law, which connects the Iron mass loss, Δm, to the corrosion current, i_{corr} (equation [7.1]).

$$\Delta m = \frac{M i_{corr} t}{nF} \qquad [7.1]$$

whereby Δm is the mass loss at the anode in g and M is the iron molar mass: M=55.8 g.mol^{-1}, t the time in s, n the number of valence electrons n=2 for the formation of Fe^{2+}, F is the Faraday constant, F= 96,500 A.s.mol^{-1}, and i_{corr} is the corrosion current in A.

In the presence of an anodic area with known homogeneous dimensions (Figure 3.1), it is easy to move on from the mass loss measurement to the thickness loss, e_{corr}, or corrosion rate, V_{corr}, expressed in mm/year. Thus, in the case of a reinforcement for reinforced concrete of a nominal diameter, D, and length, L, of uniform corrosion, the anodic surface, S_a, may be determined using equation [7.2].

$$S_a = \pi D L \qquad [7.2]$$

The mass loss, Δm, is then calculated with equation [7.3],

$$\Delta m = S_a e_{corr} \rho_s \qquad [7.3]$$

whereby ρ_s = 7.86 g/cm³ is the Iron density

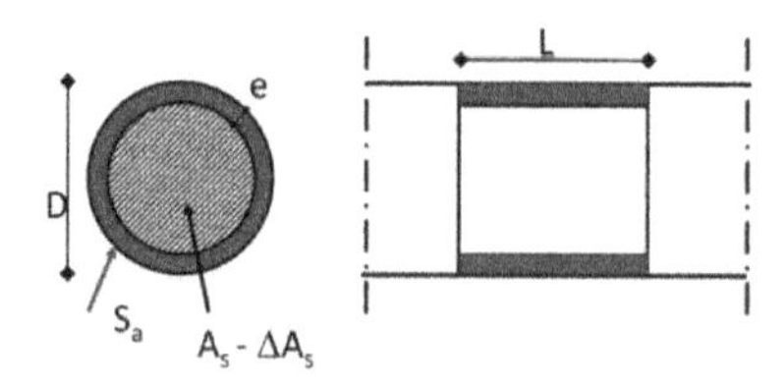

Figure 7.2. *View of homogeneous corrosion on a reinforcement. For a color version of the figure, see www.iste.co.uk/francois/corrosion.zip*

For a given corrosion kinetics there is a corresponding thickness loss (corrosion rate), whereas the reinforcement cross-section loss depends on the initial diameter. An example is given in Table 7.1 for reinforcements with a diameter of 12 mm or 16 mm.

Uniform corrosion kinetics	i_{corr} µA/cm^2	V_{corr} mm/year	ΔA_s mm^2/year	C%/year
D=12 mm	85.5	1	34.5	30.6
D=12 mm	1	0.012	0.44	0.39
D=12 mm	0.1	0.0012	0.044	0.04
D=16 mm	85.5	1	47.12	23.4
D=16 mm	1	0.012	0.59	0.29
D= 16 mm	0.1	0.0012	0.059	0.03

Table 7.1. *Uniform corrosion kinetics expressed as corrosion current density or thickness lost per year and corresponding cross-section loss for a reinforcement diameter of 12 mm and 16 mm*

In the presence of localized corrosion, we propose performing a corrosion kinetics calculation, assuming that the anodic area is based on a triangular cross-section with an opening angle of 2φ on the diameter of the reinforcement but truncated at depth e_p for a length, L (Figure 7.3). For the case of a reinforced-concrete reinforcement of a nominal diameter, D, and length, L, of uniform corrosion, the anodic surface, Sa, may be determined using equation [7.4]

$$S_a = \left(D^2 - \left(D - e_p\right)^2\right)\tan\varphi \qquad [7.4]$$

The mass loss, Δm, is then calculated with equation [7.5]

$$\Delta m = \rho_s L S_a \qquad [7.5]$$

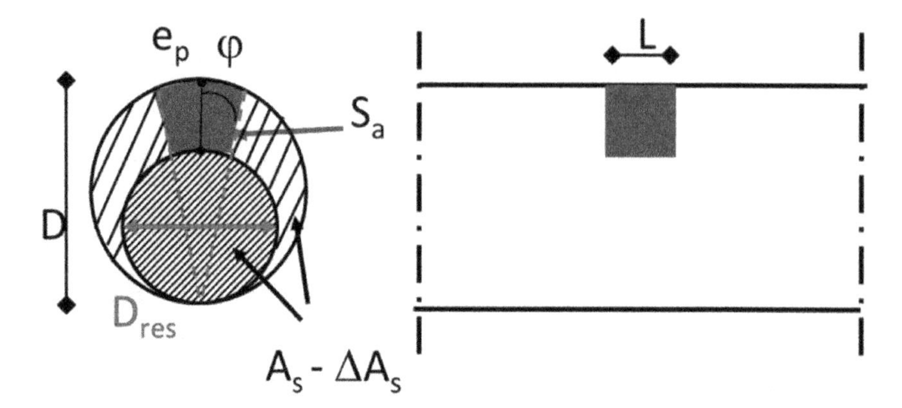

Figure 7.3. *View of localized corrosion on a reinforcement. For a color version of the figure, see www.iste.co.uk/francois/corrosion.zip*

To differentiate generalized corrosion from localized corrosion, a pitting factor is introduced that can have different definitions in the literature [AND 04a]. As such, in Chapter 3 we proposed three different pitting factors, according to whether we are interested in the geometrical appearance of the pitting for a given reinforcement cross-section (pf_g), the characterization of the difference between maximum corrosion and average corrosion on the structure (pf_s), or the difference, pf_{lc}, between localized (LC) and generalized (GC) corrosion.

Moreover, we saw in Chapter 6 that considering a residual diameter based on a pitting depth is not a reliable approach for calculating mechanical characteristics as this very much overestimates the transverse cross-section loss.

For the numerical application concerning localized corrosion kinetics, we propose to consider an angle $\varphi = \frac{\pi}{12}$, for two reinforcement diameters, 12 mm and 16 mm (Table 3.2). The pitting shape chosen of course has an impact on the corrosion current density calculation and especially on the residual cross-section calculation.

Localized corrosion kinetics	i_{corr} µA/cm^2	V_{corr} mm/year	ΔA_s mm^2/year	C%/year	C% (based on D_{res})
D=12 mm	11.1	1	6.2	5.45	16
D=12 mm	1	0.09	0.58	0.51	1.5
D=12 mm	0.1	0.009	0.06	0.05	0.15
D=16 mm	11.1	1	8.3	4.1	12.1
D=16 mm	1	0.012	0.78	0.39	1.1
D= 16 mm	0.1	0.009	0.08	0.04	0.11

Table 7.2. *Localized corrosion kinetics, expressed as corrosion current density or as pitting thickness per year and corresponding cross-section loss for a reinforcement diameter of 12 mm and 16 mm, considering either the actual cross-section loss, or the residual diameter calculated from the pit depth*

It should be noted that the localized corrosion kinetics calculated in Table 7.2 correspond to the relation proposed by Rodriguez *et al.* [ROD 96a, ROD 96b] and recalled in equation [7.6].

$$i_{corr} = \frac{D - D_{res}}{\alpha 0{,}0115t} \qquad [7.6]$$

whereby D and D_{res} (Figure 7.3) are expressed in mm, t is the corrosion duration in years, and α a pitting factor (which plays the role of the geometric pitting factor, *pfg*, defined in Chapter 3). Indeed, for a decrease in diameter of 1 mm in 1 year, equation [7.6] gives $i_{corr} = 10.9$ µA/cm^2, corresponding to the value of $i_{corr} = 11.1$ µA/cm^2 in Table 7.2.

It should also be noted that the formula by Rodriguez *et al.* [ROD 96a, ROD 96b] (equation [7.6]) enables the uniform-corrosion kinetics to be found (Table 3.1), provided that α=1 is used for the pitting factor whereas the authors define α=2 for uniform corrosion.

A prediction of a structure's service life is then possible if we have an idea of the corrosion kinetics, which is impossible to measure at present (Chapters 1–4).

7.4. Prediction of the occurrence of corrosion cracks: duration, t_c

As presented in Chapter 4, section 4.1.1, after the starting of active corrosion, the occurrence of the first corrosion cracks is obtained, which are due to the expansive nature of the corrosion products thus exerting pressure on the concrete cover.

The model to predict steel cross-section loss that has been retained in this work is that of Vidal *et al.* [VID 04], an empirical model based on the results obtained via the autopsy of beams corroded naturally for several years. This model is recalled in equation [7.7].

$$\Delta A_{s0} = A_s\left[1-\left[1-\frac{2pf_g}{\phi_0}\left(7{,}53+9{,}32\frac{c}{\phi_0}\right)10^{-3}\right]^2\right] \quad [7.7]$$

whereby ΔA_{s0} is the local cross-section loss leading to the occurrence of corrosion-induced cracks in mm², A_s is the cross-section of the reinforcement concerned (mm²), ϕ_0 is the nominal reinforcement diameter (mm), c is the distance between the reinforcement and the outer concrete surface (mm) and p_{fg} is the geometric pitting factor, taken as equal to 4 for chloride-induced corrosion and 1 for carbonation-induced corrosion.

By way of example, let us return to the comparison of the model by Vidal *et al.* with the results obtained on a naturally-corroded beam, Bs03, presented in section 4.4.1.

The calculation of the steel cross-section loss initiating cracking proposed by Vidal gives: $\Delta A_{s0} = 2.99\ \text{mm}^2$

Knowing the duration relative to the occurrence of the first corrosion cracks, t=12 months [YU 15a], it is possible to determine the corrosion kinetics corresponding to a localized cross-section loss of $\Delta A_{s0} = 2.99\ \text{mm}^2$, according to the diagram in Figure 7.3. A kinetics value of $11\ \mu\text{A/cm}^2$ is obtained (corresponding to a pitting-depth value of 1 mm/year).

It should be noted that this kinetics value is much higher than those found in the literature [AND 04b] because the latter do not take into account the localized nature of the corrosion.

7.5. Prediction of the growth in corrosion cracks in relation to the corrosion duration

Based on a summary of the ageing during a period of 28 years of the beams cast in 1984 in Toulouse [FRA 94a, FRA 94b], Zhu *et al.* [ZHU 16a] and [ZHU 16b] demonstrate that corrosion crack opening is a continuous process throughout the service life of corroded elements. In other words, this means that between the occurrence of the first corrosion cracks and the delamination of the reinforcements for corrosion crack openings reaching 3.5 to 4 mm [ZHU 16a, ZHU 16b], there is a quasi-proportionality between the maximum corrosion crack opening measured on the surface and the corrosion process duration (see Figure 7.4).

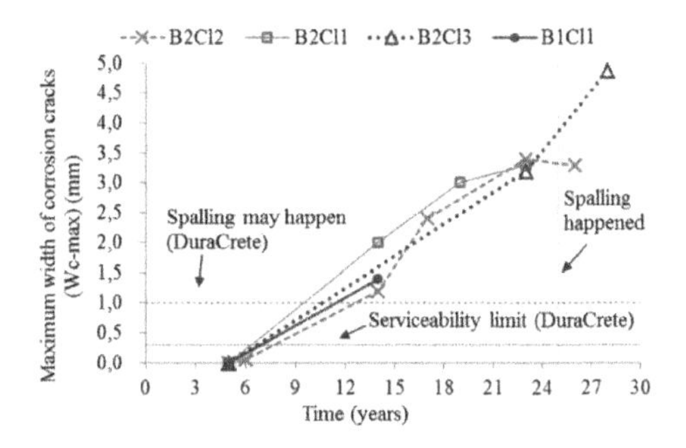

Figure 7.4. *Evolution of the maximum corrosion crack opening with respect to the corrosion duration, according to Zhu* et al. *[ZHU 16a, ZHU 16b]. For a color version of the figure, see www.iste.co.uk/francois/corrosion.zip*

It should be noted that this result is also highlighted by Vu *et al.* [VU 05] and Mullard and Stewart [MUL 11], but based on works conducted under accelerated corrosion conditions under electrical field, which is not representative of actual corrosion (as recalled in Chapter 1), but unfortunately still continues to be widely employed in current research.

The maximum corrosion crack opening is not the only visible parameter that is a function of the corrosion process duration; it also appears that the ratio of corroded reinforcement length with respect to total reinforcement length (here this is the reinforcement length facing the concrete surface), also evolves with respect to the corrosion process duration: very low at the start owing to the non-uniform nature of the corrosion of steel reinforcements in reinforced concrete, reaching a final value of 1 owing to the development of corrosion cracks along the reinforcements, which enables the corrosion to generalize all along the surface [ZHU 17b] (see Figure 7.5).

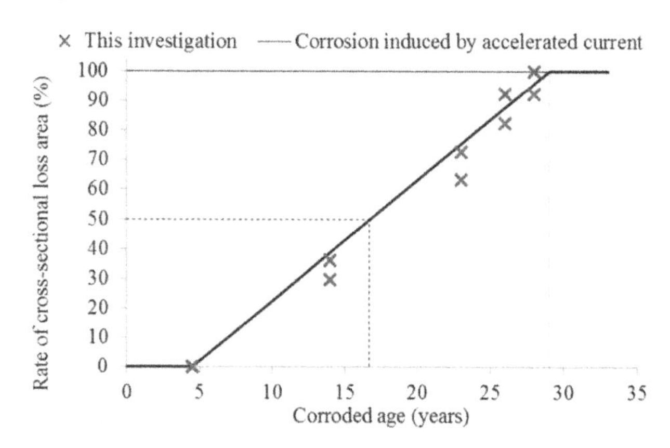

Figure 7.5. *Evolution of the ratio between the corroded reinforcement length and the reinforcement length facing the concrete surface, according to Zhu* et al. *[ZHU 17b]. For a color version of the figure, see www.iste.co.uk/francois/corrosion.zip*

The consequences of the continuous evolution in corrosion crack opening with respect to the corrosion duration as well as that of the length of corroded reinforcement are numerous.

It is thus possible to predict the evolution in corrosion rate based on two readings (as a minimum), but no doubt more accurately with three readings of maximum crack openings spaced a minimum of one year of aging apart. An illustration is proposed, concerning two geometrically identical beams, C3 and C4, aged for three and a half years under service load in artificial saline environment (cycles of two days of salt-spray spraying and two weeks of natural drying) but with different exposure: upper face cracked for C4 and lower face cracked for C3. It can therefore be expected that there will be higher corrosion kinetics for C4 than for C3, according to Yu e*t al.* [YU 16]. The tendency presented in Figure 7.6 confirms the evolution of maximum corrosion crack opening with respect to the aging time.

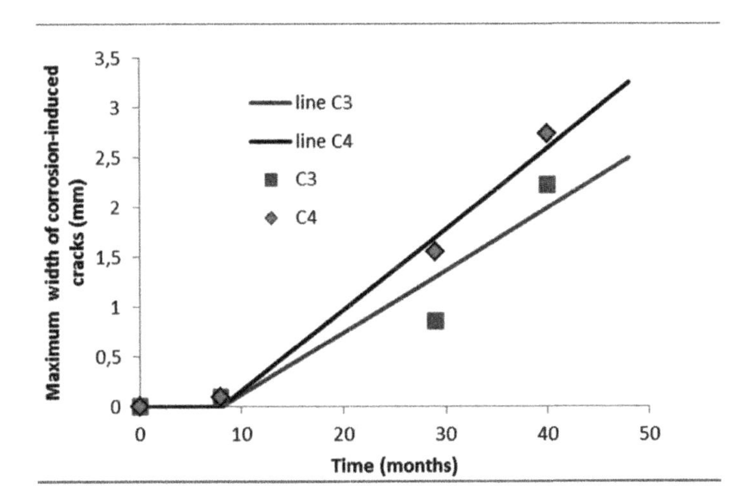

Figure 7.6. *Evolution of maximum corrosion crack opening with respect to the corrosion duration for two reinforced-concrete beams, C3 and C4. For a color version of the figure, see www.iste.co.uk/francois/corrosion.zip*

It is also possible to predict the end of service life of a structure where this is based on a crack opening criterion such as 0.3 mm for a visible crack criterion, or 1 mm to avoid the risk of delamination, in accordance with the definition proposed by DuraCrete [DUR 00].

7.6. Service-life prediction

During the propagation phase, there is therefore an increase in corrosion-induced cracking, both in terms of the growth of their opening and in terms of the extension along the reinforcements (section 7.5). The linear nature of the evolution in corrosion crack opening with respect to time is difficult to explain; indeed, after the corrosion starting in localized form, the development of the corrosion cracking leads to a generalization of the corrosion along the cracks, which thus pursue their development along the structure. Thus, two corrosion processes coexist: localized corrosion plus generalized corrosion.

In order to predict service life, the end of the structure's service needs to be defined. Limit states are therefore needed: service limit states (SLS) or ultimate limit states (ULS).

7.6.1. *Service Limit States*

The service limit time of a structure, t_{sls}, presented in Figure 7.1, can correspond to several definitions: esthetic criterion, delamination risk or excessive deflections.

7.6.1.1. *Visual SLS criterion*

The most restrictive definition corresponds to a visual criterion based on those existing in the design of structures: thus, proposals can be found that vary between 0.15 mm and 0.4 mm [AND 93, ACI 01], bearing in mind that a "visible" opening of 0.3 mm based on the design standards [ACI 05, EUR 92] may also be proposed. Nevertheless, these values are low and may be debatable as an esthetic criterion; as such, Sakai *et al.* [SAK 99] propose an opening limit of 0.8 mm.

7.6.1.2. *SLS criterion for the delamination risk*

If the owner-builder accepts the presence of visible corrosion cracks, the next stage in the corrosion process leading to end of service life of the structure is the risk of delamination of the concrete, which poses a risk to individuals' safety. The DuraCrete report [DUR 00] proposes an opening limit for corrosion-induced cracks of 1 mm; this is a very conservative value. Rodriguez *et al.* [ROD 96b] propose 2 mm and Zhu *et al.* demonstrate [ZHU 16a, ZHU 16b] that delamination only occurs for openings well above 3 mm.

7.6.1.3. *SLS criterion for the risk of excessive deflections in service*

The development of corrosion leads to a decrease in steel-concrete bond and thus to an increase in deflections under service loading. According to Zhang *et al.* [ZHA 09a] the decrease in steel-concrete bond may be quantified by a damage variable, D_c (see Chapter 6), which directly affects the bending stiffness. The criterion for end of service life proposed by Zhang *et al.* is thus the reaching of the maximum damage, $D_c=1$, corresponding to a total lack of bond between the corroded steel reinforcement and the concrete.

7.6.1.4. *Practical meaning of the SLS criteria*

Using the model by Vidal *et al.* [VID 04], it is possible to visualize, for a reinforcement, the evolution in local cross-section (LC) loss, expressed as a % of the initial cross-section, as a function of the corrosion crack openings measured on the surface of the concrete and to deduce there from the cross-section loss corresponding to the different service limit states relative to the crack opening (Figure 7.7). The reinforcement diameter is of great importance as the model by Vidal *et al.* connects a cross-section loss to an opening: a reinforcement of a larger diameter will present a lower relative loss in cross-section.

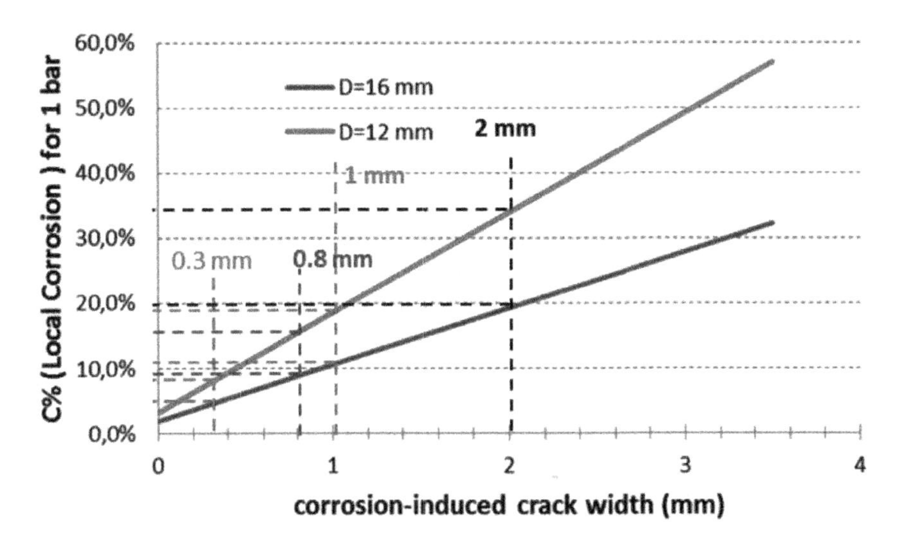

Figure 7.7. *View of the cross-section loss of a reinforcement with respect to the corrosion crack opening, generated on the surface of the concrete for two different diameters, 12 mm and 16 mm, for the different SLSs. For a color version of the figure, see www.iste.co.uk/francois/corrosion.zip*

In the presence of several tensile reinforcements, the cross-section loss on the beam cross-section will be less than that of a single bar used as an example in Figure 7.7. However, it can be observed that an opening limit of 2 mm for the delamination risk leads to a local cross-section loss of more than 20% for a diameter of 16 mm and more than 33% for a diameter of 12 mm. These values correspond to a corrosion propagation duration of 16 years and 20 years respectively, based on an identical kinetics for the two beams, of 3.25 $\mu A/cm^2$, value proposed by Yu *et al.* [YU 15a] to describe the corrosion process for beams A (D=16 mm) and B (D=12 mm), kept for almost 30 years in saline environment [FRA 94a] (Figures 7.8 and 7.9).

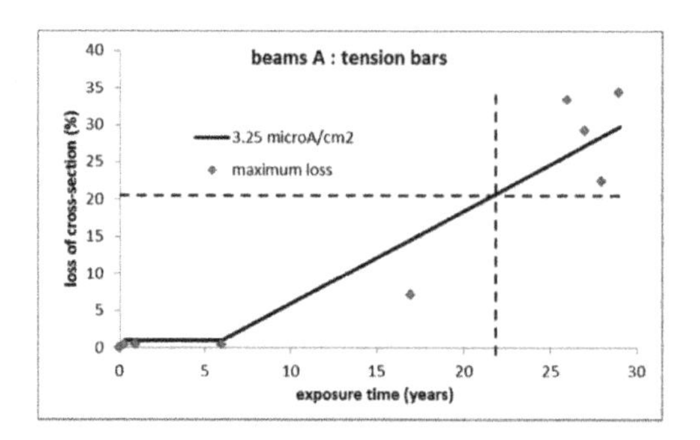

Figure 7.8. *Corrosion process expressed in terms of maximum local diameter loss with respect to time, described by a local corrosion kinetics of 3.25 µA/cm², according to Yu* et al. *[YU 15a] (beams A with tensile reinforcements 16 mm in diameter). For a color version of the figure, see www.iste.co.uk/francois/corrosion.zip*

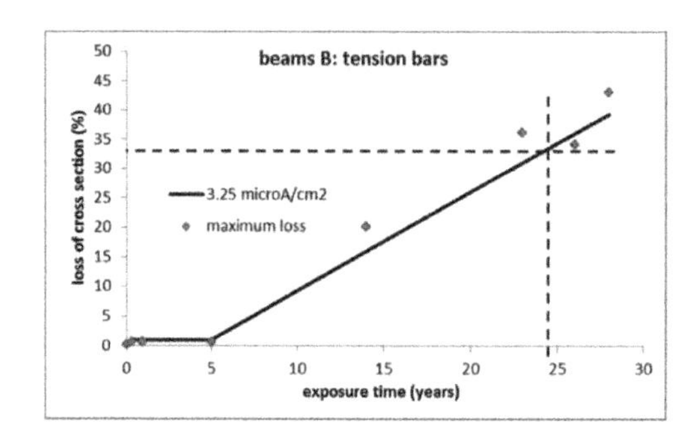

Figure 7.9. *Corrosion process expressed in terms of maximum local diameter loss with respect to time, described by a local corrosion kinetics of 3.25 µA/cm², according to Yu* et al. *[YU 15a] (beams B with tensile reinforcements 12 mm in diameter). For a color version of the figure, see www.iste.co.uk/francois/corrosion.zip*

As regards the criterion of excessive deflections, the progressive deterioration in bond leads to an increase in the deflections or a decrease in the bending stiffness: when the bond is entirely lost, only the reduction in generalized (GC) tensile steel cross-section has a limited influence on deflection, as shown in Figure 7.10, taken from the works of Zhu *et al.* [ZHU 16a]. As planned for by the SLS criterion of Zhang *et al.* [ZHA 09a], the end of service life corresponds to the end of the influence of the steel-concrete bond on bending stiffness (17 years in Figure 7.10), because beyond this point, there are no more visible signs of the continuation of the corrosion process, which is not safe. For this criterion a service life of 17 years is thus obtained.

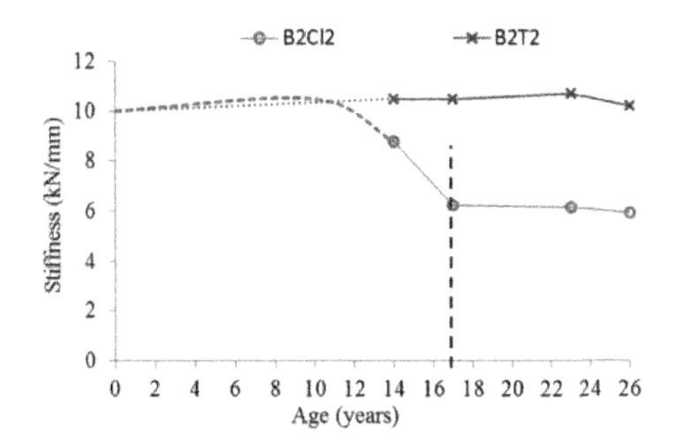

Figure 7.10. *Evolution in bending stiffness with respect to time and deterioration in the steel-concrete bond, according to Zhu* et al. *[ZHU 16a] (beams B with tensile reinforcements 12 mm in diameter). For a color version of the figure, see www.iste.co.uk/francois/corrosion.zip*

7.6.2. *Ultimate Limit States*

The service life of a structure, t_{uls}, presented in Figure 3.1, can correspond to several definitions: excessive loss of bearing capacity, limitation of the formation of the plastic hinge, etc.

7.6.2.1. *ULS criterion for bearing capacity*

Neither in the literature nor in the state of the art are there requirements to be found concerning an acceptable value of the bearing capacity loss of a structure due to corrosion, nevertheless the value of 10% cross-section loss for tensile reinforcements is often cited [CAI 03], corresponding to a decrease in the equivalent bearing capacity under bending. In this case there can then be significantly greater local losses on one of the bars making up the tensile reinforcement layer.

It should be noted that a value of 10% is low compared to the partial safety coefficients used at the design stage and also with respect to the fact that the steel cross-sections calculated are lower than those actually put in place in order to take account of the bar cross-sections actually available.

7.6.2.2. *ULS criterion for the capacity of large yielding strains (as in seismic areas)*

Neither in the literature nor in the state of the art are there requirements to be found concerning an acceptable value of reinforcement cross-section loss with respect to the reinforcement brittleness risk. Given the significant reduction in elongation at failure of the reinforcements in the presence of pits, a local cross-section loss of 10% on a bar would seem a reasonable target value. This criterion may lead to a relatively low reinforcement cross-section loss on the beam cross-section if several bars are not yet corroded. This criterion is therefore to be reserved for seismic risk and for areas where a moment redistribution is expected.

7.6.2.3. *Practical meaning of the ULS criteria*

The phenomenology of the ULS criteria is simpler than that of the SLS criteria as there is directly a quasi-proportionality between the localized cross-section loss at a beam cross-section and the bearing capacity. What is not clear, however, in the absence of recalculation recommendations for old structures, is the role of the partial safety factors used at the design stage. Indeed, if the same methodology is retained for the recalculation of the structure damaged by corrosion as that used at the design stage, a local loss of 10% proposed as an ULS criterion will lead to a shorter service life than that resulting from the SLS criteria, which is paradoxical and does not take into account the reservation of capacity induced by the rules for the basis of design. Thus, Figure 7.11, taken from the works of Zhu *et al.* [ZHU 16a], shows that an actual bearing capacity (without taking into account partial safety factors), which would be equal to the design value at ULS, is only obtained beyond 45 years of corrosion with local kinetics of 3.25 μA/cm^2.

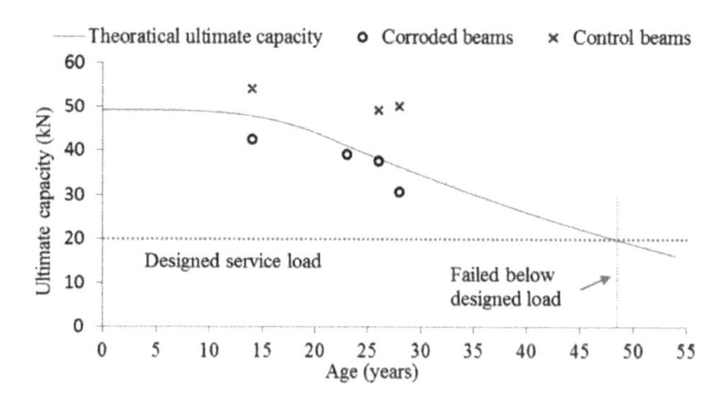

Figure 7.11. *Evolution of the actual bearing capacity with respect to the corrosion duration, and comparison with the design value at ULS, taking into account the partial safety factors, according to Zhu* et al. *[ZHU 16a] (beams B with tensile reinforcements 12 mm in diameter). For a color version of the figure, see www.iste.co.uk/francois/corrosion.zip*

7.6.3. *Predicting service life*

The results presented in this chapter and in this book demonstrate that the corrosion propagation duration is much higher than that leading to the starting of corrosion. It is therefore indispensable to consider both the corrosion initiation and the propagation in order to predict the service life of a reinforced-concrete structure, whether in the case of the estimation of the residual service life of an existing structure, or in the case of a design based on a service-life prediction (100 years, for example).

NOTE.– In the case of a diagnosis of existing structures, it will be possible to predict the service life based on the analysis of the corrosion crack openings measured on the concrete surface. Given the proportionality between the corrosion crack openings and time (see section 7.5), performing a minimum of two crack opening measurements one year apart will enable their evolution with respect to time to be predicted and the different SLS criteria presented in section 7.6.1 to be determined, as well as the ULS criteria, according to the number of bars making up the tensile reinforcements. This diagnosis may then be refined by conducting a third measurement at a later stage.

In the case of the prediction of the service life of a structure at the time it is designed, it is a little early at this stage to propose prediction methods both for the starting of corrosion owing to the uncertainties regarding initiation conditions, in particular regarding the concept of a depassivation threshold in the presence of chlorides, but also regarding the propagation duration, which is strongly influenced by the conditions at the steel-concrete interface. Nevertheless, a pragmatic approach would be to consider at the very least that the propagation duration is equal to the duration of the initiation phase.

8

Repair and Maintenance of Reinforced-Concrete Structures

8.1. Introduction

The aim of this chapter is to present the repair and/or maintenance possibilities that can actually act on the corrosion process of structures undergoing deterioration.

The conventional techniques, consisting of replacing all portions of existing concrete that are unbonded owing to corrosion with a purge of the concretes and cleaning or indeed additions of new steel cross-section to compensate for the cross-section losses due to corrosion will not be presented here. The requirements relating to conventional repairs on reinforced-concrete structures with respect to damage linked to corrosion are set out in Standard NF EN 1504 [STA 06]. Despite this recommendations context, any partial repair of a structure undergoing corrosion inevitably leads to the corrosion process being reinforced outside of non-repaired areas. This perverse or parasitic effect is due to the non-uniform nature of reinforced-concrete corrosion, widely discussed in this book. Indeed, any partial repair leads to a repassivation of the steel that was in active corrosion phase, and therefore to an increase in its electrochemical potential: these areas then become cathodic areas favoring and increasing the corrosion kinetics of the non-repaired adjacent areas. This phenomenon is known as incipient anodes in reinforced concrete repairs [CHR 16]. As a result, increasing numbers of conventional repair works are accompanied by a galvanic-type local cathodic protection in order to also sustain the areas adjacent to the repair.

In parallel to the conventional repair techniques, electrochemical maintenance of reinforced-concrete structures has considerably taken off over the last 20 years. The main maintenance techniques have been developed with a view to directly or indirectly acting on the corrosion of steel in reinforced concrete:

– Prevention and/or cathodic protection of structures against corrosion: direct action on the reinforcement layout.

– Realkalinization: restoring of the high *pH*, in case of carbonated concrete.

– Dechlorination: drawing chloride ions out of contaminated concrete.

This chapter will focus essentially on the cathodic protection technique, which is widely used in the sector of maintenance and preservation of reinforced-concrete structures at high risk of corrosion.

8.2. Prevention and cathodic protection

8.2.1. *General principals*

The history of cathodic protection began in the field of ship building at the start of the 19th Century [DAV 24]. Having for a long time been confined to marine and buried metal structures, the technology really began to develop within the sector of reinforced-concrete constructions in the late 20th Century.

In schematic terms, a metal is a crystalline (structured) assembly of cations (positive ions), the cohesion of which is assured by the metallic bonding. The metallic bonding is generated by sharing the electrons of the external layer of metal atoms to form a cloud (or gas), into which the crystalline network is plunged. Corrosion of a metal is effective when it is subject to anodic polarization with respect to its reversible equilibrium potential. At atomic scale, this polarization, consisting of removing electrons from the metal, is then reflected by a destabilization of the metallic bonding and the facilitated passage of the cations into the electrolyte. In order to inhibit the corrosion phenomenon, cathodic protection then simply consists of continually providing electrons to the reinforcement layout in order to sustain the metallic bonding. The electrons supplied favor a reduction reaction at the interface between passive steel and concrete and, according to the intensity of the protection current, limit or even eliminate the oxidation reaction on active sites. The impressed protection current then provokes an overall cathodic polarization of the corrosion system and a drop in the steel potential field values. In the cathodic protection system, the steel-reinforcement layout thus plays the role of cathode, the anode used depending on the type of procedure.

Two types of cathodic protection procedures can be differentiated between:

– Galvanic cathodic protection based on sacrificial anodes.

– Cathodic protection by impressed current using inert anodes and a current rectifier.

Under galvanic cathodic protection, the electrons are supplied to the reinforcements by means of an electrical coupling with a less noble metal, referred to as a sacrificial metal. A natural galvanic cell is thus favored, in which the steel reinforcements and the sacrificial metal constitute the system cathode and anode, respectively. With this type of protection system, it is difficult to predict the protection current released by the sacrificial anodes, often reflected by an oversizing of the installation (fairly dense anode network).

Under cathodic protection by impressed current, the supply of electrons is assured and controlled by an external current rectifier (Figure 8.1). To this end an inert auxiliary electrode (often in activated titanium) is used as a protection system anode.

The degree of the protection's effectiveness is directly linked to the intensity of the applied polarization current I_p. If the current is too low, the corrosion kinetics is simply reduced (Figure 8.1). The polarization is limited, however, due to the risk of the production of hydrogen at the cathode (reinforcement).

The choice of a protection system depends on numerous technical and financial considerations. It is considered, for example, that impressed-current systems need to be favored as soon as the extension of the structure's service life exceeds 10 years [WIL 13].

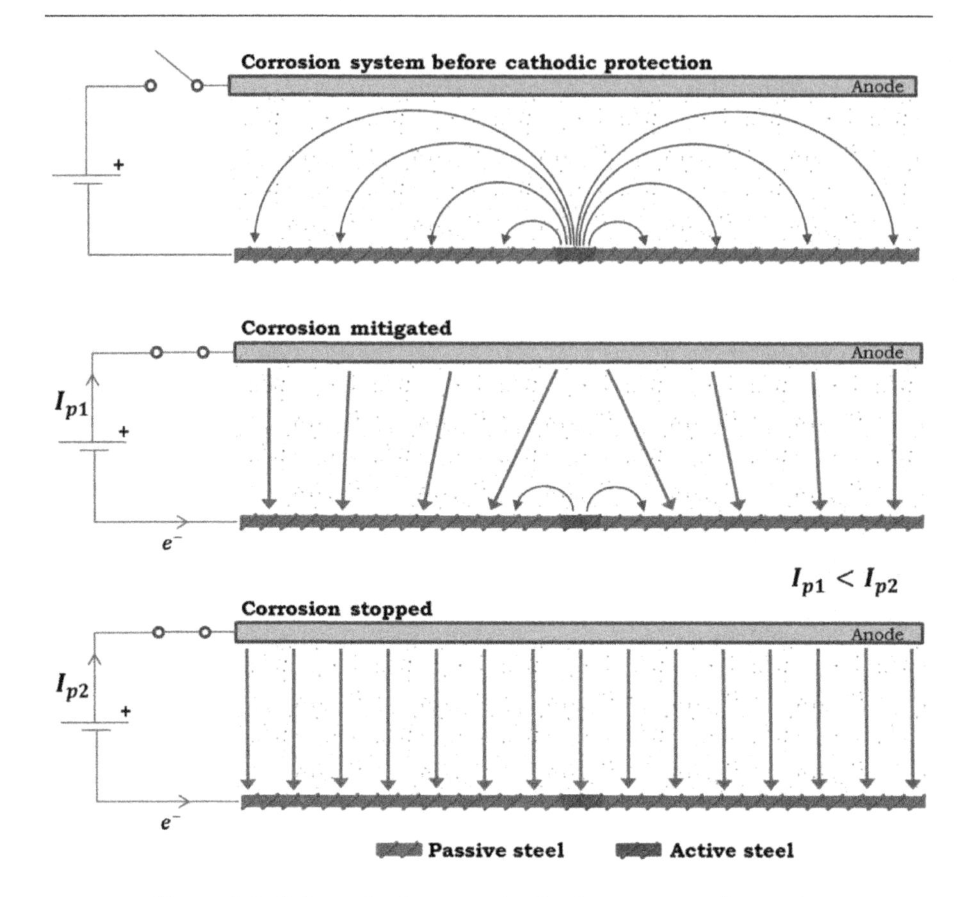

Figure 8.1. *Schematic diagram of cathodic protection. For a color version of the figure, see www.iste.co.uk/francois/corrosion.zip*

According to the cathodic polarization current level, the prevention and cathodic protection are distinguished between (Table 8.1). Cathodic prevention is implemented upstream of initiation of the corrosion phenomenon on new or recent structures presenting a proven risk. Cathodic protection concerns old structures presenting a proven corrosion state. It is to be noted that the protection level is measured in ampere per square meter of steel surface to be protected.

Standard	Rehabilitation	New Construction
European Standard EN 12696	2 mA to 20 mA/m^2	0.2 mA to 20 mA/m^2
Australian Standard AS 2832-5-2002	2 mA to 20 mA/m^2	0.2 mA to 20 mA/m^2
British Standard BS 7361	5 mA to 20 mA/m^2	N/A
NACE Standard RP0290-2000	N/A	N/A
Japanese JIS	N/A	N/A
Aramco Standard SAES-X-800	2 mA to 20 mA/m^2	N/A
Royal Commission section 16645	N/A	2 mA/m^2
SABIC Standard B01-E04	20 mA/m^2	5 mA/m^2

Table 8.1. *Protection and cathodic-prevention current ranges [CHE 14]*

8.2.2. *Standards-approach context*

Within the reinforced-concrete construction sector, the design, installation and verification of a cathodic protection system in Europe are governed by Standard EN ISO 12696 [STA 12]. Concerning the assessment of the performance of the system on site, it is necessary to switch the system to open circuit (current off) and follow the depolarization of the steel over time (Figure 8.2). The standard stipulates that the protection is effective if one of the following three criteria is applied as a minimum:

– An instantaneous "off" potential below −720 mV with respect to the Ag/AgCl/0.5M KCl electrode, the "off" potential being defined as the cut-current potential.

– A depolarization above or equal to +100 mV with respect to the "off" potential over a period of 24 hours.

– A depolarization above or equal to +150 mV with respect to the "off" potential over a period of more than 24 hours subject to a continuing decay.

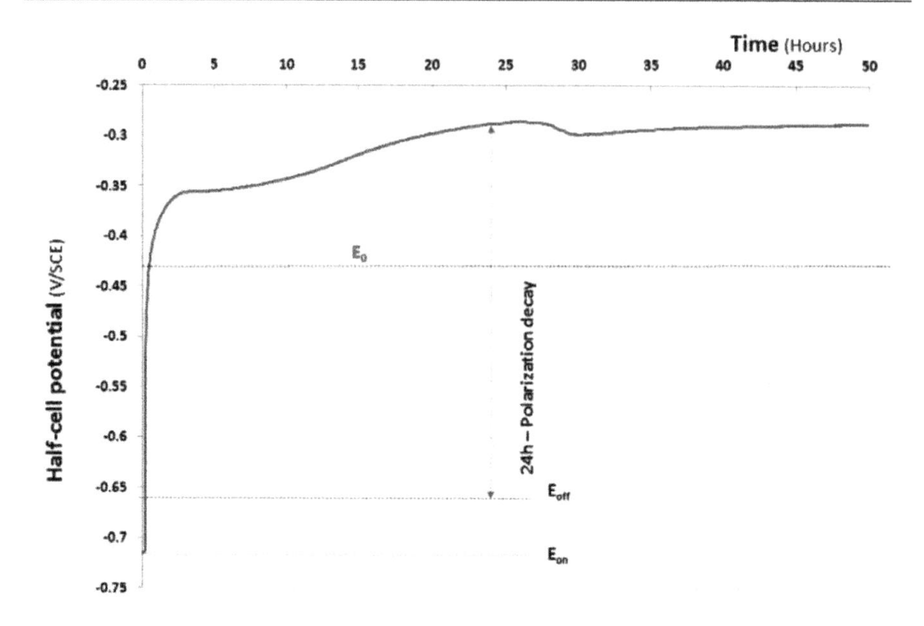

Figure 8.2. *Example of depolarization monitoring [SAS 16]. For a color version of the figure, see www.iste.co.uk/francois/corrosion.zip*

These criteria, and notably the potential decay of +100 mV, have been drawn up empirically [HAN 89].

8.2.3. *Discussion*

As mentioned above, the implementation of cathodic protection in the concrete sector is the purpose of International Standard EN ISO 12696. However, the requirements highlighted in this recommendations document are based on a very empirical, certainly overly-simplified approach to the problem.

According to an all-too-common mode of thought within the scientific community concerned, applying a cathodic protection current to a galvanic corrosion system established between active steel sites and passive steel sites would tend to make the potential field uniform at the surface of the reinforcement layout (Figure 8.3). Physically speaking, there is nothing to justify this scenario, given the non-reducible 3D nature of the problem.

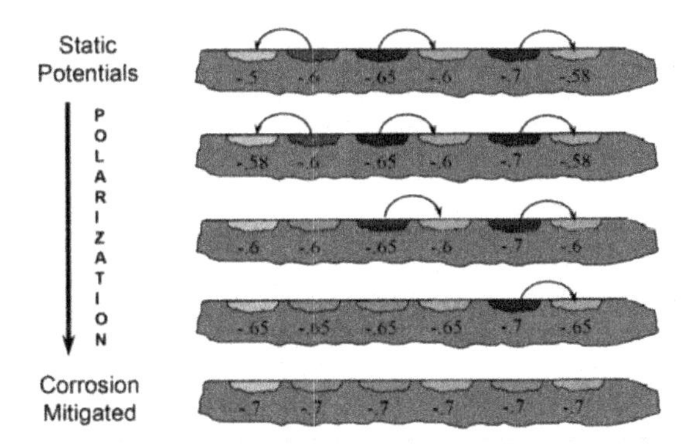

Figure 8.3. *Illustration of the principle of cathodic protection according to [NAC 90]*

Figures 8.4 and 8.5 highlight the more complex behavior of a corrosion system under cathodic protection [SAS 16]. Figure 8.4 presents a free potential map (no impressed current) collected on a reinforced-concrete wall affected by macrocell corrosion. The corrosion site is clearly identified by the significant drop in potential at the center of the image. The existence of a high potential gradient attests to a localized (or galvanic) corrosion condition between a localized active site at the center of the wall and the rest of the passive reinforcements.

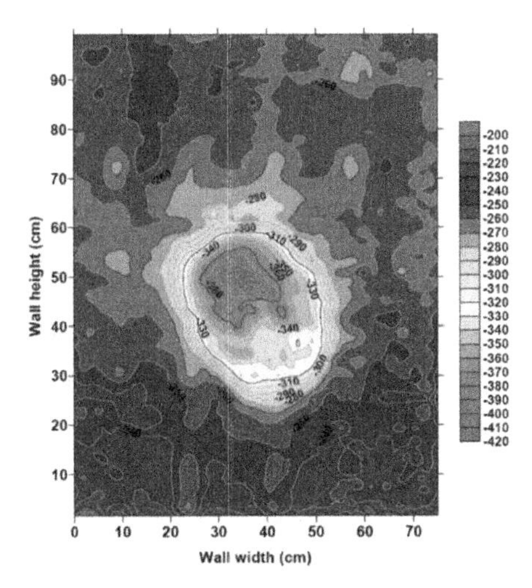

Figure 8.4. *Free potential map collected at the surface of the concrete wall (mV with respect to the saturated calomel electrode) [SAS 16]. For a color version of the figure, see www.iste.co.uk/francois/corrosion.zip*

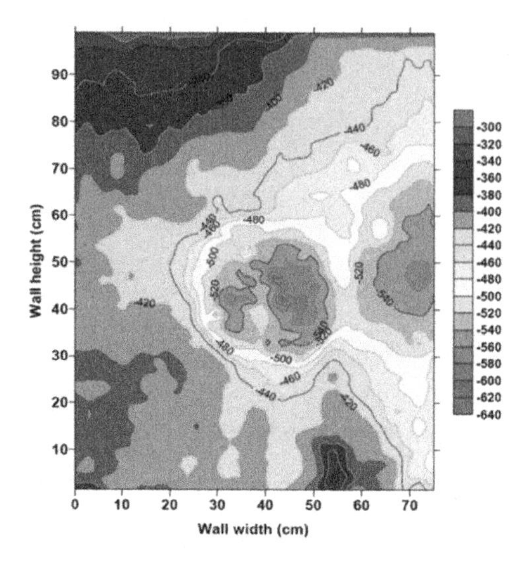

Figure 8.5. *Potential map collected at the surface of the reinforced-concrete wall under cathodic protection (mV with respect to the saturated calomel electrode) [SAS 16]. For a color version of the figure, see www.iste.co.uk/francois/corrosion.zip*

Figure 8.5 illustrates the potential map collected on this same wall under cathodic protection. Upon comparing with the free potential map, the overall decrease in the potential values induced by cathodic polarization can be observed. The active site, however, is always clearly distinct from the rest of the steel-reinforcement layout. It can even be noted that the range of the potential values is higher under cathodic protection: from −200 to −420 mV for the free corrosion system and from −300 to −640 mV, for the system under cathodic protection. This simple, reproducible experiment demonstrates that it is impossible to make the potential field uniform at the steel-reinforcement layout surface. Generally speaking, in the presence of a localized corrosion system, the distribution of a polarization, anodic or cathodic current is never uniform [LAU 16, MAR 16].

While the general principle of cathodic protection technology is not be called into question, it is necessary to combat a certain number of common preconceived ideas within the community concerning cathodic protection, which jeopardize the smooth, sustainable development of the technology. Indeed, the overall process of design, implementation and verification can be significantly improved and optimized through better comprehension of the physical phenomena at stake:

– Better knowledge of the electrochemical behavior of all components of the system (anodes, active steel and passive steel).

– Taking into consideration of the 3D nature of the physical problem (not reducible to an equivalent 1D problem) and the specific aspects of the structure to be protected (geometry, steel density, resistivity field, availability of O_2).

– Development of numerical design assistance tools.

– Proposal of more relevant performance criteria.

– Etc.

These few items constitute the pathways for scientific development of cathodic protection currently being studied within the reinforced-concrete sector.

8.3. Realkalinization and Dechlorination

Realkalinization and removal of chlorides under electrical fields are two complementary electrochemical maintenance techniques of cathodic protection. The protocols relating to the implementation of realkalinization and chloride-extraction techniques are described respectively in Standard NF EN 14038 [STA 16].

Realkalinization, required in the case of carbonated concretes, aims to raise the level of pH of the interstitial solution through the formation of hydroxyl ions, OH^-, near the reinforcements. To this end, realkalinization technology forces the dioxygen reduction reaction at the surface of the steel reinforcements, thus resulting in a cathodic polarization.

The removal of the chlorides, or dechlorination, consists of generating an electrical field, $\vec{E}$ (force field), which pushes the Cl^- ions towards the concrete surface, thus distancing them from the reinforcement layout. As the chlorides are negative charges, they are subjected to an electrical force opposite the direction of the electrical field. An electrical circuit can then be created, in which the negative charges circulate from the steel to the outside in the electrolytic portion of the circuit (ions in concrete) and from the outside to the steel in the metallic portion (electrons).

Consequently, although the objectives aimed for are different, the three electrochemical maintenance techniques described in this chapter strictly correspond to the same electrical circuit and lead to a cathodic polarization of the reinforcement layout (Figures 8.1 and 8.6). In reality, the differences are limited to:

– The levels of current involved from one technique to another.

– The protocols for polarization current application.

– The nature of the anodes used.

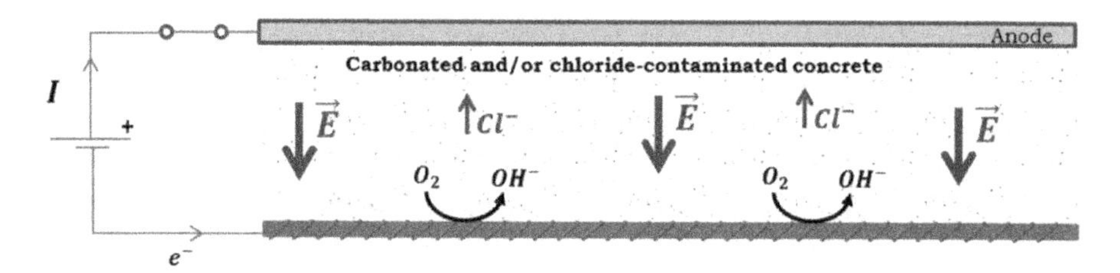

Figure 8.6. *Illustration of the electrical circuit used for realkalinization and for dechlorination. For a color version of the figure, see www.iste.co.uk/francois/corrosion.zip*

Appendix

Experimental Basis

A.1. Introduction

This appendix presents a summary of the experimental program used as reference or calculation elements in this book dedicated to corrosion.

A.2. Experimental basis

In 1984, a long-term experimental program was put in place in the INSA research laboratory, Toulouse. *Thirty-six beams* were cast, and were separated into two groups, A and B, according to the value of the cover concrete. These beams were loaded and exposed to a salt spray and the development of corrosion was monitored at different time periods. It was observed that the development of corrosion was not influenced by cracks induced by the bending load, which is contrary to the conclusions drawn by other researchers. The possible explanations were attributed to the healing of the cracks by corrosion products and to the quality of the steel-concrete interface, which depends on the casting direction.

It was also observed that for the beam of group A, cracks induced by corrosion occurred only on the tensile part, and not on the compressed part free of service cracks. This result would seem to contradict the previous conclusion affirming that the cracks induced by loading do not have any effect on the corrosion propagation. However, in the case of the A beams, the tensile bars were placed on top of the reinforcement layout whilst the compression bars were placed on the bottom. Certain defects induced by bleeding, settlement and segregation of fresh concrete occurred under the upper (tensile) bars, and thus caused the development of corrosion.

In order to clarify the role of the casting defects induced on the bars with respect to the casting direction on the development of corrosion, *twelve new beams* were set by Dang in 2010, which were named As and Bs. These beams have the same configurations as the old beams, but the position of the reinforcements has been inverted with respect to beams A and B: the tensile reinforcements of beams Bs are at the top of the formwork and those of beam A at the bottom of the formwork. Nevertheless, in this experimental program, the casting defects are mixed with other factors, such as cover concrete depth, the exposure conditions and so on, and it is difficult to highlight solely the effect of casting defects. Consequently, in 2013, *four beams* with a unique cover concrete of 20 mm, were cast in order to study the effect of the casting defects alone, and identified under group C.

For the three experimental programs beginning in 1984, 2010 and 2013, respectively, all beams were loaded and kept in an aggressive environment generated by wetting-drying cycles with a salt fog of 35 g/l NaCl (see Figure A.1).

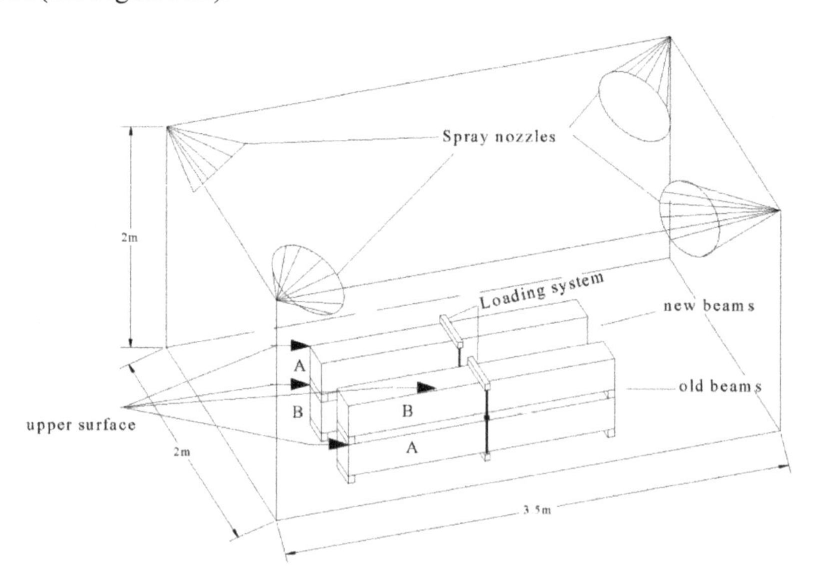

Figure A.1. *Diagram of the sustaining under load of the reinforced-concrete beams of type A, B, As, Bs and C in a conservation chamber with salt spray*

Figure A.2 shows the differences in exposure experienced by the beams in the salt-fog chamber: in particular, the upper face of the beams remains wet for longer, resulting in a greater chloride concentration on the surface. Moreover, the system of loading by coupling two beams gives a tensile lower surface for one of the beams and a compressed upper surface for the other beam.

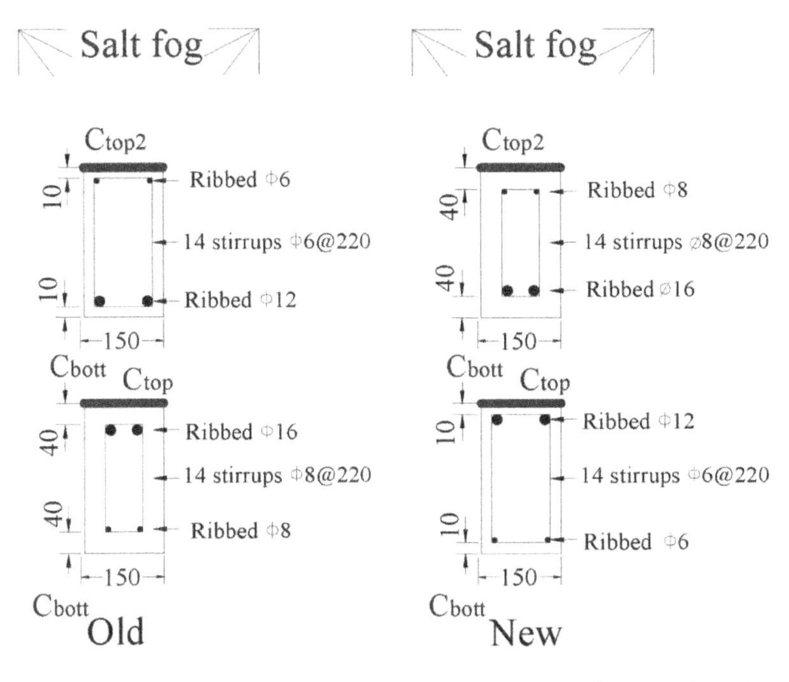

Figure A.2. *View of the differences in environment in the salt-fog chamber: the upper face of the beams remains wet for longer, and the chloride concentration on the surface is therefore higher. For a color version of the figure, see www.iste.co.uk/francois/corrosion.zip*

The different beams tested, and the results of which were used in this book, are listed in Table A.1. Details of the tests and results can be found in the following references: [CAS 00a, CAS 00b, CAS 03, DAN 12, DAN 13, DAN 14, KHA 14, FRA 06a, FRA 06b, VID 04, YU 15a, YU 15b, YU 15c, YU 16, ZHA 09a, ZHA 09b, ZHA 10, ZHU 13, ZHU 14, ZHU 15a, ZHU 15b, ZHU 16a, ZHU 16b, ZHU 17a, ZHU 17b].

Beam	Setting year	Type	Corrosion duration	Loading, kN.m
Bs03	2010	corroded	19 m	13.5
Bs04	2010	corroded	27 m	21.2
Bs02	2010	corroded	36 m	13.5
As06	2010	corroded	19 m	13.5
C1	2013	corroded	29 m	21.2
C2	2013	corroded	29 m	21.2
C3	2013	corroded	29 m	13.5
C4	2013	corroded	29 m	13.5
B1Cl1	1984	corroded	14 years	13.5
B2Cl1	1984	corroded	23 years	21.2
B2T	1984	control	26 years	21.2
B2T3	1984	control	28 years	21.2
B2Cl2	1984	corroded	26 years	21.2
B2Cl3	1984	corroded	28 years	21.2
A2T	1984	control	27 years	21.2
A1T	1984	control	27 years	13.5
A2Cl1	1984	corroded	27 years	21.2
A2Cl2	1984	corroded	27 years	21.2
A1Cl1	1984	corroded	17 years	13.5
A2Cl3	1984	corroded	26 years	21.2

Table A.1. *Summary of beams whose results were used in the various chapters*

Bibliography

[ACI 01] ACI 224R01, *Control of Cracking in Concrete Structures*, American Concrete Institute, Farmington Hills, Mich, 2001.

[ACI 05] ACI 318-05, *Building Code Requirements for Structural Concrete*, American Concrete Institute, Farmington Hills, Mich, 2005.

[ALA 09] AL-AHMAD S., TOUMI A., VERDIER J. *et al.*, "Effect of crack opening on carbon dioxide penetration in cracked mortar samples", *Materials and Structures*, vol. 42, pp. 559–566, 2009.

[ALE 12] ALEXANDER M., OTIENO M., BEUSHAUSEN H., "Corrosion in cracked concrete and related issues of structural durability and service life", *Conférence NUCPERF 2012*, Cadarache, France, November 2012.

[ALH 11] AL-HARTHY A.S., STEWART M.G., MULLARD J., "Concrete cover cracking caused by steel reinforcement corrosion", *Magazine of Concrete Research*, vol. 63, no. 9, pp. 655–667, 2011.

[ALO 98] ALONSO C., ANDRADE C., RODRIGUEZ J. *et al.*, "Factors controlling cracking of concrete affected by reinforcement corrosion", *Materials and Structures*, vol. 31, pp. 435–441, 1998.

[ALW 90] ALWIS W.A.M., "Trilinear moment–curvature relationship for reinforced concrete beams", *ACI Structural Journal*, vol. 87, no. 3, pp. 276–283, 1990.

[AND 93] ANDRADE C., ALONSO C., MOLINA F.J., "Cover cracking as a function of rebar corrosion: Part 1—experimental test", *Material and Structures*, vol. 26, pp. 453–464, 1993.

[AND 01] ANDRADE C., KEDDAM M., NOVOA X.R. *et al.*, "Electrochemical behaviour of steel rebars in concrete: influence of environmental factors and cement chemistry", *Electrochimica Acta*, vol. 46, nos 24–25, pp. 3905–3912, 2001.

[AND 04] ANDRADE C., ALONSO C., GULIKERS J. *et al.*, "Test methods for on-site corrosion rate measurement of steel reinforcement in concrete by means of the polarization resistance method", *Materials and Structures*, vol. 37, pp. 623–643, 2004.

[ANG 09] ANGST U.M., ELSENER B., LARSEN C.K. *et al.*, "Critical chloride content in reinforced concrete – a review", *Cement and Concrete Research*, vol. 39, no. 12, pp. 1122–1138, 2009.

[ANG 15] ANGST U.M., BÜCHLER M., "On the applicability of the Stern–Geary relationship to determine instantaneous corrosion rates in macro-cell corrosion", *Materials and Corrosion*, vol. 66, no. 10, pp. 1017–1028, 2015.

[ANG 17a] ANGST U.M., ELSENER B., "Chloride threshold values in concrete – a look back and ahead", *ACI Special Publication*, "Chloride Limits and Thresholds for Concrete Containing Supplementary Cementitious Materials (SCMs)", 2017.

[ANG 17b] ANGST U.M., GEIKER M.R., ALEXANDER M. *et al.*, "The steel–concrete interface", *Materials and Structures*, vol. 50, no. 2, p. 143, 2017.

[ARY 95] ARYA C., WOOD L.A., The relevance of cracking in concrete to corrosion of reinforcement, The Concrete Society, Camberley, Surrey, Technical Report no. 44, 1995.

[AS 09] AS3600-2009, Australian Standard for Concrete Structures, 2009.

[AST 09] ASTM C876-09, *Standard Test Method for Corrosion Potentials of Uncoated Reinforcing Steel in Concrete*, ASTM International, West Conshohocken, PA, available at: www.astm.org, 2009.

[BAE 82] BAEL, "Béton Armé aux Etats Limites", 1982.

[BAL 16] BALESTRA C.E.T., LIMA M., SILVA A.R. *et al.*, "Corrosion degree effect on nominal and effective strengths of naturally corroded reinforcement", *Journal of Materials in Civil Engineering*, vol. 28, no. 10, p. 04016103, 2016.

[BAL 18] BALAYSSAC J.P., GARNIER V., *Non-destructive Testing and Evaluation of Civil Engineering Structures*, ISTE Press, London and Elsevier, Oxford, 2018.

[BAZ 80] BAZANT Z.P., GAMBAROVA P., "Rough cracks in reinforced concrete", *Journal of the Structural Division*, vol. 106, pp. 819–842, 1980.

[BEE 78] BEEBY A.W., "Corrosion on reinforcing steel in concrete and its relation to cracking", *The Structural Engineer*, vol. 56, no. 3, pp. 77–81, 1978.

[BEE 83] BEEBY A.W., "Cracking cover at corrosion of reinforcement", *Concrete International*, vol. 5, no. 2, pp. 35–40, 1983.

[BEN 97] BENTUR A., DIAMOND S., BERKE N., *Steel Corrosion in Concrete, Fundamentals and Civil Engineering Practice*, E & FN Spon, London, pp. 41–43, 1997.

[BIS 05] BISCHOFF P.H., "Reevaluation of deflection prediction for concrete beams reinforced with steel and fiber-reinforced polymer bars", *Journal of Structural Engineering*, vol. 131, no. 5, pp. 752–767, 2005.

[CAI 03] CAIRNS J., DU Y., LAW D.W., "Structural assessment of corrosion-damaged bridges", *Proceedings of the Conference on Structural Faults and Repair*, London, Engineering Technics Press, Edinburgh, 2003.

[CAR 86] CARREIRA J.D., CHU K., "The moment–curvature relationship of reinforced concrete members", *ACI Structural Journal,* vol. 83, no. 2, pp. 191–198, 1986.

[CAS 00a] CASTEL A., FRANÇOIS R., ARLIGUIE G., "Mechanical behaviour of corroded reinforced concrete beams – Part 1: experimental study of corroded beams", *Materials and Structures*, vol. 33, pp. 539–544, 2000.

[CAS 00b] CASTEL A., FRANÇOIS R., ARLIGUIE G., "Mechanical behaviour of corroded reinforced concrete beams—Part 2: bond and notch effects", *Materials and Structures*, vol. 33, no. 9, pp. 545–551, 2000.

[CAS 03] CASTEL A., VIDAL T., FRANÇOIS R. *et al.*, "Influence of steel–concrete interface quality on reinforcement corrosion induced by chlorides", *Magazine of Concrete Research*, vol. 55, no. 2, pp. 151–159, 2003.

[CAS 11] CASTEL A., FRANÇOIS R., "Modeling of steel and concrete strains between primary cracks for the prediction of cover-controlled cracking in RC beams", *Engineering Structures*, vol. 33, no. 12, pp. 3668–3675, 2011.

[CAS 16] CASTEL A., KHAN I., FRANÇOIS R. *et al.*, "Modelling steel concrete bond strength reduction due to corrosion", *ACI Structural Journal*, vol. 5, no. 113, pp. 973–982, 2016.

[CEB 89] CEB DESIGN GUIDE, *Durable Concrete Structures – CEB Design Guide*, 2nd ed., fib Bulletins no. 182, 1989.

[CEB 99] CEB-FIP MODEL CODE, *Structural Concrete. Basis of Design*, vol. 2, Updated Knowledge of the CEP-FIP Model Code 1990, July 1999.

[CHA 92] CHAN H.C., CHUENG Y.K., HUANG Y.P., "Crack analysis of reinforced concrete tension members", *Journal of Structural Engineering*, vol. 118, no. 8, pp. 2118–2132, 1992.

[CHA 18] CHALHOUB C., FRANÇOIS R., CARCASSÉS M., "A new approach to determine the chloride threshold initiating corrosion: preliminary results", *Proceedings of the ICCRRR 2018 Conference*, Cape Town, October 2018.

[CHE 14] CHESS P., CREVELLO G., NOYCE P., "Anode performance: the use of ballasted mixed metal oxide coated titanium anodes in impressed current systems installed within historic steel frame masonry clad structures", *Proceedings of Concrete Solutions, 5th International Conference on Concrete Repair*, Belfast, September 2014.

[CHI 05] CHITTY W.-J., DILLMANN P., L'HOSTIS V. *et al.*, "Long-term corrosion resistance of metallic reinforcements in concrete—a study of corrosion mechanisms based on archaeological artefacts", *Corrosion Science*, vol. 47, no. 6, pp. 1555–1581, 2005.

[CHR 16] CHRISTODOULOU C., GOODIER C., GLASS G. *et al.*, "Incipient anodes in reinforced concrete repairs: a cause or a consequence?", in BEUSHAUSEN H. (ed.), *Performance-Based Approaches for Concrete Structures: Proceedings*, Cape Town, South Africa, 21–23 November, 2016.

[CLÉ 12] CLÉMENT A., LAURENS S., ARLIGUIE G. *et al.*, "Numerical study of the linear polarisation resistance technique applied to reinforced concrete for corrosion assessment", *European Journal of Environmental and Civil Engineering*, vol. 16, pp. 491–504, 2012.

[COM 85] COMITE EURO-INTERNATIONAL DU BETON, Design Manual on Cracking and Deformations, Ecole Polytechnique Fédérale de Lausanne, 1985.

[DAN 12] DANG V.H., FRANÇOIS R., L'HOSTIS V., "Effects of cracks on both initiation and propagation of re-bar corrosion due to carbonation", *NUCPERF*, Cadarache, 2012.

[DAN 13] DANG V.H., FRANÇOIS R., "Influence of long-term corrosion in chloride environment on mechanical behaviour of RC beam", *Engineering Structures*, vol. 48, pp. 558–568, 2013.

[DAN 14] DANG V.H., FRANÇOIS R., "Prediction of ductility factor of corroded reinforced concrete beams exposed to long-term aging in chloride environment", *Cement and Concrete Composites*, vol. 53, pp. 136–147, 2014.

[DAR 85] DARWIN D., "Debate: crack width, cover, and corrosion", *Concrete International*, vol. 7, no. 5, pp. 20–32, 1985.

[DAV 24] DAVY H., "On the corrosion of copper sheeting by seawater, and on methods of preventing this effect, and on their application to ships of war and other ships", *Proceedings of the Royal Society*, vol. 114, pp. 151–158, 1824.

[DUR 00] DURACRETE, The European Union-Brite EuRam III, DuraCrete, Final Technical report, Document BE95-1347/R17, 2000.

[ELM 05] EL MAADDAWY T., SOUDKI K., TOPPER T., "Long-term performance of corrosion-damaged reinforced concrete beams", *ACI Structural Journal*, vol. 102, pp. 649–656, 2005.

[ELM 07] EL MAADDAWY T., SOUDKI K., "A model for prediction of time from corrosion initiation to corrosion cracking", *Cement and Concrete Composites*, vol. 29, no. 3, pp. 168–175, 2007.

[ELC 03] ELCHALAKANI M., ZHAO X.L., GRZEBIETA R., "Tests of cold-formed circular tubular braces under cyclic axial loading", *Journal of Structural Engineering*, vol. 129, no. 4, pp. 507–514, 2003.

[ELS 02] ELSENER B., "Macrocell corrosion of steel in concrete – implications for corrosion monitoring", *Cement & Concrete Composites*, vol. 24, pp. 65–72, 2002.

[ELS 03] ELSENER B., ANDRADE C., GULIKERS J. *et al.*, "Hall-cell potential measurements—potential mapping on reinforced concrete structures", *Materials and Structures*, vol. 36, no. 7, pp. 461–471, 2003.

[EUR 92] EUROCODE 2: NF EN 1992-1-1: "Design of concrete structures, Part I. General rules and rules for buildings", 1992.

[FAN 04] FANG C., LUNDGREN K., CHEN L. *et al.*, "Corrosion influence on bond in reinforced concrete", *Cement and Concrete Research*, vol. 34, pp. 2159–2167, 2004.

[FLO 82] FLOEGL H., HERBERT H., MANG A., "Tension stiffening concept on bond slip", *Journal of Structural Engineering*, vol. 108, no. 12, pp. 2681–2701, 1982.

[FRA 88] FRANÇOIS R., MASO J.C., "Effect of damage in reinforced concrete on carbonation or chloride penetration", *Cement and Concrete Research*, vol. 18, pp. 961–970, 1988.

[FRA 94a] FRANÇOIS R., ARLIGUIE G., MASO J.C., "Durabilité du béton armé soumis à l'action des chlorures", *Annales de l'I.T.B.T.P.*, vol. 529, pp. 1–48, 1994.

[FRA 94b] FRANÇOIS R., ARLIGUIE G., "Durability of loaded reinforced concrete in chloride environment", *ACI Special Publication*, vol. 145, pp. 573–596, 1994.

[FRA 94c] FRANÇOIS R., ARLIGUIE G., BARDY D., "Electrode potential measurements of concrete reinforcement for corrosion evaluation", *Cement and Concrete Research*, vol. 24, no. 3, pp. 401–412, 1994.

[FRA 99] FRANÇOIS R., ARLIGUIE G., "Effect of microcracking and cracking on the development of corrosion in reinforced concrete members", *Magazine of Concrete Research*, vol. 51, no. 2, pp. 143–150, 1999.

[FRA 06a] FRANÇOIS R., CASTEL A., VIDAL T. *et al.*, "Long term corrosion behavior of reinforced concrete structures in chloride environment", *Journal de Physique IV*, vol. 136, pp. 258–293, 2006.

[FRA 06b] FRANÇOIS R., CASTEL A., VIDAL T., "A finite macro-element for corroded reinforced concrete", *Materials and Structures*, vol. 5, no. 39, pp. 569–582, 2006.

[GEN 97] GENIN J.M.R., REFAIT P., RAHARINAIVO A., "Green rusts, intermediate corrosion products formed on rebar in concrete in the presence of carbonation or chloride ingress", *International Conference Understanding Corrosion Mechanisms of Metals in Concrete: A Key to Improving Infrastructure Durability*, Cambridge, USA, 1997.

[GHA 17a] GHANTOUS R.M., POYET S., L'HOSTIS V. *et al.*, "Effect of accelerated carbonation conditions on the characterization of load-induced damage in reinforced concrete members", *Materials and Structures*, vol. 50, no. 175, pp. 1–10, 2017.

[GHA 17b] GHANTOUS R.M., POYET S., L'HOSTIS V. *et al.*, "Effect of crack openings on carbonation-induced corrosion", *Cement and Concrete Research*, vol. 95, pp. 257–269, 2017.

[GIL 99] GILBERT R.I., "Flexural crack control for reinforced concrete beams and slabs: an evaluation of design procedures", *ACMSM 16, Proceedings of the 16th Conference on the Mechanics of Structures and Materials*, Sydney, pp. 175–180, 1999.

[GOT 71] GOTO Y., "Cracks formed in concrete around deformed tension bars", *Journal of the American Concrete Institute*, vol. 68, pp. 244–251, 1971.

[GUL 99] GULIKERS J., "Numerical simulation of corrosion rate determination by linear polarization", in MESINA C., ANDRADE C., ALONSO J. *et al.* (eds), *Rilem PRO 18, Workshop on Measurement and Interpretation of On-Site Corrosion Rate*, Madrid, February 1999.

[GUP 89] GUPTA A.K., MAESTRINI S.R., "Post-cracking behavior of membrane reinforced concrete elements including tension-stiffening", *Journal of Structural Engineering*, vol. 115, no. 4, pp. 957–976, 1989.

[HAN 89] HAN M.K., Cathodic protection of concrete bridge components, Strategic Highway Research Program, Progress Report for Contract SHRP-87-C-102B, 1989.

[HEL 12] HELLAND S., "Design for service life: implementation of Model Code 2010 rules in the operational, code ISO 16204", *Structural Concrete*, vol. 14, pp. 10–18, 2012.

[HOR 07] HORNE A.T., RICHARDSON J.G., BRYDSON R.M.D., "Quantitative analysis of the microstructure of interfaces in steel reinforced concrete", *Cement and Concrete Research*, vol. 37, pp. 1613–1623, 2007.

[HUE 05] HUET B., L'HOSTIS V., MISERQUE F. *et al.*, "Electrochemical behaviour of mild steel in concrete: influence of pH and carbonate content of concrete pore solution", *Electrochimica Acta*, vol. 51, no. 1, pp. 172–180, 2005.

[ISM 08] ISMAIL M., TOUMI A., FRANÇOIS R. *et al.*, "Effect of crack opening on local diffusion of chloride in cracked mortar samples", *Cement and Concrete Research*, vol. 38, pp. 1106–1111, 2008.

[IZQ 04] IZQUIERDO D., ALONSO C., ANDRADE C. *et al.*, "Potentiostatic chloride threshold values for rebar depassivation: experimental and statistical study", *Electrochimica Acta*, vol. 49, nos 17–18, pp. 2731–2739, 2004.

[JEN 09] JENKINS D., *Prediction of Cracking and Defections*, International Codes Provisions and Recent Research, Concrete Solution, Concrete Institute of Australia, 2009.

[KEN 15] KENNY A., KATZ A., "Statistical relationship between mix properties and the interfacial transition zone around embedded rebar", *Cement and Concrete Composites*, vol. 60, pp. 82–91, 2015.

[KHA 14] KHAN I., CASTEL A., FRANÇOIS R., "Prediction of reinforcement corrosion using crack width in corroded reinforced concrete beams", *Cement and Concrete Research*, vol. 56, pp. 84–96, 2014.

[KRA 01] KRANC S.C., SAGÜES A.A., "Detailed modeling of corrosion macrocells on steel reinforcing in concrete", *Corrosion Science*, vol. 43, pp. 1355–1372, 2001.

[LAU 16] LAURENS S., HENOCQ P., ROULEAU N. *et al.*, "Steady-state polarization response of chloride-induced macrocell corrosion systems in reinforced concrete – numerical and experimental investigations", *Cement and Concrete Research*, vol. 79, pp. 272–290, 2016.

[LEO 77] LEONHARDT F., "Crack control in concrete structures", IABSE Surveys NoS4/77, International Association for Bridge and Structural Engineering, 1977.

[LUN 05] LUNDGREN K., "Bond between ribbed bars and concrete. Part 2: the effect of corrosion", *Magazine of Concrete Research*, vol. 57, no. 7, pp. 383–395, 2005.

[LUN 15] LUNDGREN K., TAHERSHAMSI M., ZANDI K. *et al.*, "Tests on anchorage of naturally corroded reinforcement in concrete", *Materials and Structures*, vol. 48, no. 7, pp. 2009–2022, 2015.

[MAL 10] MALUMBELA G., ALEXANDER M., MOYO P., "Variation of steel loss and its effect on the ultimate flexural capacity of RC beams corroded and repaired under load", *Construction and Building Materials*, vol. 24, pp. 1051–1059, 2010.

[MAR 01] MARCOTTE T.D., Characterization of chloride-induced corrosion products that form in steel-reinforced cementitious materials, PhD Thesis, University of Waterloo, 2001.

[MAR 16] MARCHAND J., LAURENS S., PROTIERE Y. *et al.*, "A numerical study of polarization tests applied to corrosion in reinforced concrete", *ACI Special Publication*, vol. 312, pp. 1–12, 2016.

[MIC 13] MICHEL A., SOLGAARD A.O.S., PEASE B.J. *et al.*, "Experimental investigation of the relation between damage at the concrete–steel interface and initiation of reinforcement corrosion in plain and fibre reinforced concrete", *Corrosion Science*, vol. 77, pp. 308–321, 2013.

[MON 03] MONTEMOR M., SIMOES A., FERREIRA M., "Chloride-induced corrosion on reinforcing steel: from the fundamentals to the monitoring techniques", *Cement and Concrete Composites*, vol. 25, pp. 491–502, 2003.

[MOR 02] MORRIS W., VICO A., VASQUEZ M. *et al.*, "Corrosion of reinforcing steel evaluated by means of concrete resistivity measurements", *Corrosion Science*, vol. 44, pp. 81–99, 2002.

[MUL 11] MULLARD J., STEWART M.G., "Corrosion-induced cover cracking: new test data and predictive models", *ACI Structural Journal*, vol. 108, no. 1, pp. 71–79, 2011.

[NAC 90] NACE STANDARD RP0290-90, Standard recommended practice, Item no. 53072, April 1990.

[NAS 10] NASSER A., CLEMENT A., LAURENS S. *et al.*, "Influence of steel–concrete interface condition on galvanic corrosion currents in carbonated concrete", *Corrosion Science*, vol. 52, no. 9, pp. 2878–2890, 2010.

[NEV 95] NEVILLE A., "Chloride attack of reinforced concrete — an overview", *Materials and Structures*, vol. 28, pp. 63–70, 1995.

[OES 97] OESTERLE R.G., The role of concrete cover in crack control criteria and corrosion protection, RD serial no. 2054, Portland Cement Association Skokie, IL, 1997.

[OTI 10] OTIENO M.B., ALEXANDER M., BEUSHAUSEN H.D., "Corrosion in cracked and uncracked concrete – influence of crack width, concrete quality and crack reopening", *Magazine of Concrete Research*, vol. 62, no. 6, pp. 393–404, 2010.

[OU 16] OU Y.C., SUSANTO Y.T.T., ROH H., "Tensile behavior of naturally and artificially corroded steel bars", *Construction and Building Materials*, vol. 103, pp. 93–104, 2016.

[PET 96] PETTERSSON K., JORGENSEN O., "The effect of cracks on reinforcement corrosion in high-performance concrete in a marine environment", *Proceedings of the 3rd ACI/CANMET International Conference on the Performance of Concrete in the Marine Environment*, St Andrews-by-the-Sea, Canada, pp. 185–200, 1996.

[PIY 04] PIYASENA R., LOO Y.C., FRAGOMENI S., "Factors influencing spacing and width of cracks in reinforced concrete; new prediction formulae", *Advances in Structural Engineering*, vol. 7, no. 1, pp. 49–60, 2004.

[POU 05] POUPARD O., L'HOSTIS V., LAURENS S. *et al.*, "*Benchmark des poutres de la Rance*: Damage diagnosis of reinforced concrete beams after 40 years exposure in marine environment by non destructive tools", *Proceedings of the EUROCORR 2005 Conference*, Lisbon, 4–8 September 2005.

[PRA 90] PRAKHYA G.K.V., MORLEY C.T., "Tension stiffening and moment–curvature relations of reinforced concrete elements", *ACI Structural Journal*, vol. 87, no. 5, pp. 597–605, 1990.

[ROD 96a] RODRIGUEZ J., ORTEGA L.M., CASAL J. *et al.*, "Corrosion of reinforcement and service life of concrete structures", *Proceedings of the 7th DBMC International Conference*, Stockholm, pp. 117–126, 1996.

[ROD 96b] RODRIGUEZ J., ORTEGA L.M., CASAL J. *et al.*, "Assessing structural conditions of concrete structures with corroded reinforcement", *4th International Congress on Concrete in Service Mankind*, Dundee, UK, 1996.

[SAG 92] SAGÜES A.A., KRANC S.C., "On the determination of polarization diagrams of reinforcing steel in concrete", *Corrosion*, vol. 48, pp. 624–633, 1992.

[SAJ 15] SAJEDI S., HUANG Q., "Probabilistic prediction model for average bond strength at steel–concrete interface considering corrosion effect", *Engineering Structures*, vol. 99, pp. 120–131, 2015.

[SAK 99] SAKAI K., SHIMOMURA T., SUGIYAMA T., "Design of concrete structures in the 21st century", *Controlling Concrete Degradation*, Dundee, Scotland, pp. 28–44, 1999.

[SAL 96] SALEEM M., SHAMEEM M., HUSSAIN S. *et al.*, "Effect of moisture, chloride and sulphate contamination on the electrical resistivity of Portland cement concrete", *Construction and Building Materials*, vol. 10, pp. 209–214, 1996.

[SAS 16] SASSINE E., LAURENS S., FRANÇOIS R. *et al.*, "Experiments and numerical simulations on impressed current cathodic protection technique applied to steel reinforced concrete structures", *Proceedings of the EUROCORR 2016*, Montpellier, France, September 2016.

[SAS 17] SASSINE E., LAURENS S., FRANÇOIS R. *et al.*, "A critical discussion on some common beliefs in the field of half-cell potential measurements in steel reinforced concrete", *Materials and Structures*, submitted, 2017.

[SCH 97] SCHIESSL P., RAUPACH M., "Laboratory studies and calculations on the influence of crack width on chloride-induced corrosion of steel in concrete", *ACI Materials Journal*, vol. 94, pp. 56–62, 1997.

[SCO 07] SCOTT A.N., ALEXANDER M., "The influence of binder type, cracking and cover on corrosion rates of steel in chloride-contaminated concrete", *Magazine of Concrete Research*, vol. 59, no. 7, pp. 495–505, 2007.

[SOH 15] SOHAIL M.G., LAURENS S., DEBY F. *et al.*, "Significance of macrocell corrosion of reinforcing steel in partially carbonated concrete: numerical and experimental investigation", *Materials and Structures*, vol. 48, pp. 217–233, 2015.

[SOM 81] SOMAYAJI S., SHAH S.P., "Bond stress versus slip relationship and cracking response of tension members", *ACI Journal*, vol. 78, no. 3, pp. 217–225, 1981.

[SÖY 03] SÖYLEV T.A., FRANÇOIS R., "Quality of steel–concrete interface and corrosion of reinforcing steel", *Cement and Concrete Research*, vol. 33, no. 9, pp. 1407–1415, 2003.

[STA 06] STANDARD NF EN 1504-7, "Produits et systèmes pour la protection et la réparation des structures en béton – Définitions, prescriptions, maîtrise de la qualité et évaluation de la conformité", *Part 7: Protection contre la corrosion des armatures*, AFNOR, 2006.

[STA 12] STANDARD EN 12696, "Cathodic protection of steel in concrete ISO 12696", 2012.

[STA 14] STANDARD NF EN 206, "Béton – Spécification, performance, production et conformité", 2014.

[STA 16] STANDARD NF EN 14038, "Réalcalinisation électrochimique et traitements d'extraction des chlorures applicables au béton armé", 2016.

[STE 57] STERN M., GEARY A.L., "Electrochemical polarization", *Journal of the Electrochemical Society*, vol. 104, pp. 56–63, 1957.

[STE 95] STEEHBLOW H.H., "Mechanisms of pitting corrosion", in MARCUS P., OUDAR J. (eds), *Corrosion Mechanisms in Theory and Practice*, Marcus Dekker, pp. 243–275, 1995.

[SUZ 90] SUZUKI K., OHNO Y., PRAPARNTANATORN S. *et al.*, "Mechanism of steel corrosion in cracked concrete", in PAGE C., TREADAWAY K., BRAMFORTH P. (eds), *Corrosion of Reinforcement in Concrete*, Society of Chemical Industry, London, pp. 19–28, 1990.

[TAH 14] TAHERSHAMI M., ZANDI K., LUNDGREN K. *et al.*, "Anchorage of naturally corroded bars in reinforced concrete structures", *Magazine of Concrete Research*, vol. 14, no. 66, pp. 729–744, 2014.

[TRA 16] TRAN V.Q., Contribution à la compréhension des mécanismes de dépassivation des armatures d'un béton exposé à l'eau de mer : théorie et modélisation thermochimique, Doctoral Thesis, Ecole Centrale de Nantes, 2016.

[TUU 82] TUUTTI L.K., Corrosion of Steel in Concrete, Cement and Concrete Research Institute, 1982.

[VEC 00] VECCHIO F., "Disturbed stress field model for reinforced concrete: formulation", *ASCE Journal of Structural Engineering*, vol. 126, no. 9, pp. 1070–1077, 2000.

[VID 04] VIDAL T., CASTEL A., FRANÇOIS R., "Analyzing crack width to predict corrosion in reinforced concrete", *Cement and Concrete Research*, vol. 34, pp. 165–174, 2004.

[VU 05] VU K., STEWART M.G., MULLARD J., "Corrosion-induced cracking: experimental data and predictive models", *ACI Structural Journal*, vol. 102, no. 5, pp. 719–726, 2005.

[VU 10] VU N.A., CASTEL A., FRANÇOIS R., "Response of post-tensioned concrete beams with unbonded tendon including serviceability and ultimate state", *Engineering Structures*, vol. 32, pp. 556–569, 2010.

[WAR 06] WARKUS J., RAUPACH M., GULIKERS J., "Numerical modelling of corrosion – theoretical backgrounds", *Materials and Corrosion*, vol. 57, pp. 614–617, 2006.

[WAR 10] WARKUS J., RAUPACH M., "Modelling of reinforcement corrosion – geometrical effects on macrocell corrosion", *Materials and Corrosion*, vol. 61, pp. 494–504, 2010.

[WIL 13] WILSON K., JAWED M., NGALA V., "The selection and use of cathodic protection systems for the repair of reinforced concrete structures", *Construction and Building Materials*, vol. 39, pp. 19–25, 2013.

[YU 15a] YU L., FRANÇOIS R., DANG V.H. *et al.*, "Development of chloride-induced corrosion in pre-cracked RC beams under sustained loading: effect of load-induced cracks, concrete cover, and exposure conditions", *Cement and Concrete Research*, vol. 67, pp. 246–258, 2015.

[YU 15b] YU L., FRANÇOIS R., DANG V.H. *et al.*, "Structural performance of RC beams damaged by natural corrosion under sustained loading in a chloride environment", *Engineering Structures*, vol. 96, pp. 30–40, 2015.

[YU 15c] YU L., FRANÇOIS R., DANG V.H. *et al.*, "Distribution of corrosion and pitting factor of steel in corroded RC beams", *Construction and Building Materials*, vol. 95, pp. 384–392, 2015.

[YU 16] YU L., FRANÇOIS R., GAGNÉ R., "Influence of steel–concrete interface defects induced by top-casting on development of chloride-induced corrosion in RC beams under sustained loading", *Materials and Structures*, vol. 49, pp. 5169–5181, 2016.

[ZHA 09a] ZHANG R., CASTEL A., FRANÇOIS R., "Serviceability limit state criteria based on steel–concrete bond loss for corroded reinforced concrete in chloride environment", *Materials and Structures*, vol. 42, pp. 1407–1421, 2009.

[ZHA 09b] ZHANG R., CASTEL A., FRANÇOIS R., "The corrosion pattern of reinforcement and its influence on serviceability of reinforced concrete members in chloride environment", *Cement and Concrete Research*, vol. 39, no. 11, pp. 1077–1086, 2009.

[ZHA 10] ZHANG R., CASTEL A., FRANÇOIS R., "Concrete cover cracking with reinforcement corrosion of RC beam during chloride-induced corrosion process", *Cement and Concrete Research*, vol. 40, pp. 415–425, 2010.

[ZHU 13] ZHU W., FRANÇOIS R., "Effect of corrosion pattern on the ductility of tensile reinforcement extracted from a 26-year-old corroded beam", *Advances in Concrete Construction*, vol. 2, pp. 121–137, 2013.

[ZHU 14] ZHU W., FRANÇOIS R., "Experimental investigation of the relationships between residual cross-section shapes and the ductility of corroded bars", *Construction and Building Materials*, vol. 69, pp. 335–345, 2014.

[ZHU 15a] ZHU W., FRANÇOIS R., "Structural performance of RC beams in relation with the corroded period in chloride environment", *Materials & Structures*, vol. 48, pp. 1757–1769, 2015.

[ZHU 15b] ZHU W., FRANÇOIS R., CORONELLI D. *et al.*, "Failure mode transitions of corroded deep beams exposed to marine environment for long period", *Engineering Structures*, vol. 96, pp. 66–77, 2015.

[ZHU 16a] ZHU W., FRANÇOIS R., FANG Q. *et al.*, "Influence of long-term chloride diffusion in concrete and the resulting corrosion of reinforcement on the serviceability of RC beams", *Cement and Concrete Composites*, vol. 71, pp. 144–152, 2016.

[ZHU 16b] ZHU W., FRANÇOIS R., "Prediction of the residual load-bearing capacity of naturally corroded beams using the variability of tension behaviour of corroded steel bars", *Structure and Infrastructure Engineering*, vol. 12, no. 2, pp. 143–158, 2016.

[ZHU 17a] ZHU W., FRANÇOIS R., POON C.S. *et al.*, "Influences of corrosion degree and corrosion morphology on the ductility of steel reinforcement", *Construction and Building Materials*, vol. 148, pp. 297–306, 2017.

[ZHU 17b] ZHU W., FRANÇOIS R., LIU Y., "Propagation of corrosion and corrosion patterns of bars embedded in RC beams stored in chloride environment for various periods", *Construction and Building Materials*, vol. 145, pp. 147–156, 2017.

Index

N, O, P

S, T, Y

Printed in the United States
By Bookmasters